# Lecture Notes in Computer Science 995

Edited by G. Goos, J. Hartmanis and J. van Leeuwen

Advisory Board: W. Brauer  D. Gries  J. Stoer

**Springer**
Berlin
Heidelberg
New York
Barcelona
Budapest
Hong Kong
London
Milan
Paris
Santa Clara
Singapore
Tokyo

Silvia M. Müller   Wolfgang J. Paul

# The Complexity
# of Simple Computer
# Architectures

 Springer

Series Editors

Gerhard Goos, Karlsruhe University, Germany

Juris Hartmanis, Cornell University, NY, USA

Jan van Leeuwen, Utrecht University, The Netherlands

Authors

Silvia M. Müller
Wolfgang J. Paul
Fachbereich Informatik, Universität des Saarlandes
Im Stadtwald, D-66123 Saarbrücken, Germany

Cataloging-in-Publication data applied for


Die Deutsche Bibliothek - CIP-Einheitsaufnahme

**Müller, Silvia Melitta:**
The complexity of simple computer architectures / Silvia M.
Müller ; Wolfgang J. Paul. - Berlin ; Heidelberg ; New York :
Springer, 1995
  (Lecture notes in computer science ; 995)
  ISBN 3-540-60580-0
NE: Paul, Wolfgang J.:; GT


CR Subject Classification (1991): C.1, B.1, C.4, C.5, B.2, B.7, B.6


ISBN 3-540-60580-0 Springer-Verlag Berlin Heidelberg New York

© Springer-Verlag Berlin Heidelberg 1995
Printed in Germany

Typesetting: Camera-ready by author
SPIN 10485838    06/3142 – 5 4 3 2 1 0    Printed on acid-free paper

# Preface

This book presents a formal model for evaluating the cost effectiveness of computer architectures. It is a revised version of the model from [MP90]. The model can cope with a wide range of architectures from CPU design to parallel supercomputers.

In particular, the model permits analysis of cost/time trade-offs with very reasonable effort. This does not contradict the widely held opinion that such analyses require the complete specification of designs in some technology. Technically speaking, we simply exhibit an easy way to interpret architectures presented in modern textbooks as complete designs and to derive their cost and cycle time.

In order to illustrate the procedure of trade-off analyses, we analyze quantitatively several non-pipelined design alternatives for a well known RISC architecture called DLX [HP90]. Among other things, we determine the cost and run time impact of full hardware support for precise interrupts and perform a comparison of hardwired and microcoded control. In addition, we will formally prove that our interrupt mechanism of the DLX architecture handles nested interrupts correctly.

Our analyses confirm some rules of thumb known in engineering and suggest some further maxims:

- The cost of a flipflop relative to the cost of an AND gate determines whether it is cost efficient to encode the states of a hardwired control automaton. If this ratio is smaller than 2, encoding is no option for reasonably sized automata.

- In a non-pipelined DLX design, hardwired control requires at most 13% of the total cost of the design.

- Even without fancy optimizations, the control unit can be designed such that it usually does not lie on the time critical path of a non-pipelined DLX design.

- Full hardware support for precise interrupts can be provided at 28% of the cost of the non-pipelined DLX design or less. The additional functionality does not impact the cycle time.

- Microcoded control can only compete with hardwired control when the technology provides very cheap ROMs (a ROM cell has at most 0.35 times the cost of an inverter).

- The well adapted instruction format has a much stronger impact on the cost of microcoded control than optimizing the coding of the micro-instructions.

## Acknowledgements

The authors would particularly like to acknowledge Robert Knuth for analyses and exercises concerning restricted fanout and for proof-reading the manuscript.

During Spring/Summer 1995, the second author used the manuscript as lecture notes for the Computer Architecture Course at the University of Saarland at Saarbrücken. Thanks also go to the students of that course for their feedback.

September 1995                                                        Silvia M. Müller
                                                                     Wolfgang J. Paul

# Acknowledgements

The authors would particularly like to acknowledge Robert Knuth for analyses and exercises concerning restricted fanout and for proof-reading the manuscript.

During Spring/Summer 1995, the second author used the manuscript as lecture notes for the Computer Architecture Course at the University of Saarland at Saarbrücken. Thanks also go to the students of that course for their feedback.

Silvia M. Müller
Wolfgang J. Paul

# Contents

# Contents

# Chapter 1

# Introduction

The well known textbook [HP90] represents the current state of the art concerning the use of quantitative methods in computer architecture. The approach of Hennessy and Patterson deviates from previous textbooks in several important ways:

1. Workload is formally represented by benchmark programs.

2. The central role of compilers for the performance of computer architectures is clearly identified.

3. The combined effect of hardware and compiler on system performance is shown to be amenable to systematic quantitative analysis.

A formal hardware model is not provided in [HP90]. Thus, the bulk of the quantitative analysis is concerned with the number of machine cycles, whereas cycle time and hardware cost are necessarily treated on a more qualitative level.

Also, without a formal hardware model one cannot quantify the impact of many design strategies experienced engineers use in order to improve their designs. There are, for example, many ways to reduce the delay of the critical path. That changes the cost and the cycle time of the design but usually does not impact the number of machine cycles respectively the CPI ratio. Modeling cost and delay of the hardware is therefore necessary for analyzing many kinds of trade-offs and for studying the impacts of many design strategies. In our view, those studies should be an essential *and quantitative* part of computer architecture research and lectures.

Of course, such studies are performed whenever one is actually designing a machine in a particular technology. One is easily tempted to believe, that this is the only way in which such studies can be done, but this is not necessarily so:

Rather than entering a whole design into say a schematics entry system and then letting the system compute the area of the components and the delay on the critical path, it clearly suffices to find out, what *would* be the result of this computation if we *would* enter the design into the system.

In this paper we show how to get a reasonable approximation of these numbers with very little incremental effort if we are willing to specify data paths at the level of [HP93] and control at the level of [HP90]. We proceed in the following way:

In chapter 2 we develop the formal hardware model. We start out with combinatorial circuits, but we allow different costs and propagation delays for different kinds of gates. In two steps we then augment the model to include first registers and then also RAM and tristate drivers. It turns out, that the rules which govern the enabling and disabling drivers can be captured with a very simple and intuitive formal condition.

In chapter 3, we specify a few combinatorial circuits, which are repeatedly used in almost any machine design. These include decoders, shifters as well as different kinds of adders and counters. Thus, in our framework this chapter is the equivalent of a library of macros in a CAD system. Technically we proceed in a very quick way:

1. We usually specify circuits for $n$-bit wide inputs by a recursive definition with respect to $n$.

2. From the definition we can immediately read off recurrence relations governing the cost $C(n)$ and the delay $T(n)$ of the circuits defined.

3. *We do not even bother to solve these equations.* Instead, we translate them in the most straightforward way into functions of the programming language C.

If one wants to get into hardware at all, one has to specify certain circuits anyway in some form or another. Thus no extra work is involved in the first step. The only incremental work is to read off two equations from every picture and to type them into a C program.

Note, that paper and pencil alone will not suffice in order to verify the results of this book. On the other hand, a CAD system is definitely not required (and would slow things down). A PC on which to run simple C programs is all one needs. It seems to be a reasonable assumption that any computer scientist has free and easy access to this resource.

In very comfortable design systems, hardwired control can be specified in form of a finite automaton. Chapter 4 provides this comfort in our framework. We analyze constructions for realizing a finite automaton either as a Moore automaton or as a Mealy automaton. In both cases, one can then read off formulae for cost and delay of the resulting hardware from a state diagram of the automaton in a fairly straightforward way.

In the remaining chapters of this paper we deal in a formal way with some well known machine designs. In chapter 5, our approach is applied to a MARK-I like architecture and some trade-offs analyses are performed. The purpose of that chapter is to illustrate the specification, analysis and evaluation of designs on a very simple architecture.

In chapters 6 and 7, we specify and analyze several non-pipelined designs for the DLX RISC machine. In chapter 8 we augment the control to handle interrupts. We also formalize and prove that our interrupt mechanism handles interrupts in a correct and precise manner. In chapter 9 we compare the cost effectiveness of hardwired control with the cost effectiveness of microcoded control for the DLX design.

For the designs under consideration we can formally confirm the following rules of thumb in engineering:

- If the technology provides cheap flipflops, i.e., a flipflop is at most 2 times more expensive than an AND gate, binary coding is not a realistic option for the implementation of reasonable sized automata (20 to 50 states). If on the other hand the relative cost of flipflops is high, i.e., a flipflop is at least 8 times more expensive than an AND gate, coding the states in binary definitely improves the cost.

- In a non-pipelined DLX design, hardwired control requires at most 13% of the total cost of the design.

- Even without fancy optimizations, the control unit can be designed such that it usually does not lie on the time critical path of a non-pipelined DLX design.

- Full hardware support for precise interrupts can be provided at 28% of the cost of the non-pipelined DLX design or less. For technologies with very expensive storage components this fraction is only about 14%. The additional functionality does not impact the the cycle time.

- Microcoded control can only compete with hardwired control when the technology provides very cheap ROMs (a ROM cell has at most 0.35 times the cost of an inverter).

- On a non-pipelined DLX design with microcoded control, a well adapted instruction format (respectively a well designed decode table) is essential for a microcoded control to be competitive. A naive approach may result in a huge cost increase.

# Chapter 2

# The Formal Architecture Model

For us, an architecture **A** consists of four components: the hardware **H**, the instruction set **Is**, the high level programming language **Pr** and the Compiler **Co**.

$$\mathbf{A} = (\mathbf{H}, \mathbf{Is}, \mathbf{Pr}, \mathbf{Co})$$

Since most users write their applications in programming languages and since we want to measure the performance of the whole system, we included the programming language and the compiler. Compilers have a big impact on the performance of a computer system.

The major difference over other approaches is the hardware model, because it handles cost and delays of designs. This information is essential for the evaluation and the comparison of designs. Before describing the hardware model in all detail, we introduce the performance model and the quality metrics.

## 2.1 The Performance Model

Each of our CPU designs runs at a constant clock cycle time $t_C$. The execution time of a workload can therefore be expressed as the number of clock cycles times the cycle time. Usually, it is much easier to count the instructions for a given workload and determine its average amount of clock cycles per instruction ($CPI$) than to directly count the cycles of the whole program. The execution time of a given workload can then be expressed as the product of the cycle time, the CPI ratio and the instruction count $IC$:

$$T = t_C \cdot CPI \cdot IC.$$

The $CPI$ ratio depends on the workload and on the hardware design. The execution scheme of the instruction set defines how many cycles $CPI_I$ an instruction $I$ requires on average. On the other hand, the workload together with the compiler defines an instruction count $IC_I$ for each machine instruction, and so the CPI value can be expressed by

$$CPI = \frac{\sum_{I \in \mathbf{Is}} IC_I \cdot CPI_I}{IC} = \sum_{I \in \mathbf{Is}} W_I \cdot CPI_I$$

$$W_I = \frac{IC_I}{IC}.$$

$W_I$ indicates the frequency of instruction $I$ in the workload.

## 2.2   Quality Metrics

We evaluate the designs based on a benchmark **Be** that models the workload **W** and is usually written in the high level programming language **Pr**. The compiler **Co** translates the benchmark into the machine language. The resulting assembly program then defines the instruction count and the frequencies of each instruction.

The execution time $T_A(Be)$ of the benchmark **Be** on architecture **A** is the product of the cycle time $t_C$, the $CPI(Be)$ value and the instruction count $IC(Be)$, as introduced in the previous section:

$$T_A(Be) = t_C \cdot CPI(Be) \cdot IC(Be).$$

The performance $P_A(Be)$ of architecture A is the reciprocal of the benchmark's execution time. The quality $Q_{q,A}(Be)$ ([Grü94]) is the weighted geometric mean $wgm()$ of the performance and of the reciprocal of its cost $C_A$.

$$P_A(Be) = \frac{1}{T_A(Be)}$$

$$Q_{q,A}(Be) = wgm(q, P_A(Be), \frac{1}{C_A}) = \frac{P_A(Be)^{1-q}}{C_A^q}$$

If there is no possibility for confusion, we omit the parameter $Be$ in the performance and quality formulae in order to make them more readable.

The weighting parameter $q \in [0, 1]$ determines whether cost or performance has a greater impact on the quality. Therefore, we denote $q$ as *quality parameter*. Commonly used values are:

- $q = 0$: Only performance counts, $Q = P$.

- $q = \frac{1}{2}$: The resulting quality metric $Q = \sqrt{\frac{P}{C}}$ models the cost/performance ratio.

For $q > \frac{1}{2}$, the cost is more important than the performance, but that is usually not desired.

## 2.3 Comparison of Architectures

Improving a design is an incremental process. Therefore, we are interested in how much a modification improves a given design. The following definitions and theorems are taken from [Grü94].

**Definition 2.1** *Two architectures $A$ and $B$ with cost $C_A$ (respectively $C_B$) and performance $P_A(Be)$ (respectively $P_B(Be)$) on a benchmark $Be$ can be compared using the following measures:*

$$\text{Cost ratio:} \quad C(A, B, Be) \quad := \quad \frac{C_A}{C_B}$$

$$\text{Performance ratio:} \quad P(A, B, Be) \quad := \quad \frac{P_A(Be)}{P_B(Be)}$$

$$\text{Quality ratio:} \quad Q_q(A, B, Be) \quad := \quad \frac{Q_{q,A}(Be)}{Q_{q,B}(Be)}$$

**Theorem 2.1** *The quality ratio can be computed directly from the cost ratio and the performance ratio:*

$$Q_q(A, B, Be) = wgm(q, P(A, B, Be), \frac{1}{C(A, B)}) = \frac{P(A, B, Be)^{1-q}}{C(A, B)^q}.$$

**Proof:**

$$Q_q(A, B, Be) = \frac{Q_{q,A}(Be)}{Q_{q,B}(Be)} = \frac{wgm(q, P_A(Be), C_A^{-1})}{wgm(q, P_B(Be), C_B^{-1})}$$

$$= \frac{P_A(Be)^{1-q}}{C_A^q} \cdot \frac{C_B^q}{P_B(Be)^{1-q}} = \frac{P_A(Be)^{1-q}}{P_B(Be)^{1-q}} \cdot \frac{C_B^q}{C_A^q}$$

$$= \frac{P(A, B, Be)^{1-q}}{C(A, B)^q}$$

$$\square$$

Design changes usually cause trade–offs between cost and performance. In those cases, we are also interested in the value of parameter $q$ which yields the same quality for both design variants.

**Definition 2.2** *Let $A$ and $B$ be two architectures with cost $C_A > C_B$ and performance $P_A(Be) > P_B(Be)$ on benchmark $Be$. The quality parameter $eq$ with*

$$Q_{eq}(A, B, Be) = \frac{P(A, B, Be)^{1-eq}}{C(A, B)^{eq}} = 1$$

*is called* **parameter of equal quality** *$EQ(A, B, Be)$.*

Once again, if there is no possibility for confusion, we omit the parameter $Be$.

**Theorem 2.2** *Let $C = C(A, B)$ and $P = P(A, B, Be)$. Under the premises of the previous definition, eq can be expressed as*

$$eq = \frac{\log P}{\log PC}$$

*For $q < eq$, architecture A is better than B, otherwise B is the better choice.*

**Proof:**

$$1 \; = \; Q_{eq}(A, B, Be) \; = \; \frac{P^{1-eq}}{C^{eq}}$$

$$\Leftrightarrow \qquad (P \cdot C)^{eq} \; = \; P$$

$$\Leftrightarrow \quad eq \cdot \log(PC) \; = \; \log P$$

$$\Leftrightarrow \qquad\qquad eq \; = \; \frac{\log P}{\log PC}$$

The quality ratio is defined as the weighted geometric mean of $P$ and $\frac{1}{C}$. For $q < eq$, performance is more important; thus, A is better than B. For $q > eq$, it is the other way around. $\qquad\square$

The smaller $eq$ is, the more emphasis has to be put on the performance gain of A. In this sense, $eq$ is a quantitative measure for the evaluation of cost/performance trade-offs.

We only consider the range of quality parameter $q \in [0.2, 0.5]$ as appropriate for the following reasons. Usually, more emphasis is put on the performance than on the cost, thus $q \leq 0.5$. Let two designs A and B have an equal quality parameter EQ(A, B, Be) of 0.2, and let A be two times faster than B. Their cost ratio C(A, B) then equals 16, i.e., design A is 16 times more expensive than design B. A higher cost ratio would rarely be accepted, thus $q \geq 0.2$.

## 2.4   The Hardware Model

Our hardware model is derived from switching theory, it defines meaning, cost and delay of the hardware of an architecture. For the formal definition we proceed incrementally from basic components via combinatorial circuits and clocked circuits to the general case. Only in the general case we treat components with control signals like RAMs and tristate drivers.

| Cost | Motorola | Venus |
|---|---|---|
| $C_{inv}$ | 1 | 1 |
| $C_{nand}, C_{nor}$ | 2 | 2 |
| $C_{and}, C_{or}$ | 2 | 2 |
| $C_{xor}, C_{xnor}$ | 4 | 6 |
| $C_{ff}$ | 8 | 12 |
| $C_{mux}$ | 3 | 3 |
| $C_{driv}$ | 5 | 6 |
| $C_{RAMcell}$ | 2 | 12 |
| $C_{ROMcell}$ | 0.25 | 2 |

| Delay | Motorola | Venus |
|---|---|---|
| $D_{inv}$ | 1 | 1 |
| $D_{nand}, D_{nor}$ | 1 | 1 |
| $D_{and}, D_{or}$ | 2 | 1 |
| $D_{xor}, D_{xnor}$ | 2 | 2 |
| $D_{ff}$ | 4 | 4 |
| $D_{mux}$ | 2 | 2 |
| $D_{driv}$ | 2 | 1 |
| $R_1$ | 3 | 1.5 |
| $R_2$ | 10 | 0 |

Table 2.1: Cost and delay of the basic components; ff stands for a flipflop

## 2.4.1 Basic Components

In a first step, we define cost and delay of all basic components, like gates, flipflop, RAM, with the standard meaning. Their cost is measured in *gate equivalents* [g] and their propagation delays in *gate delays* [d]. These values are normalized to the cost (delay) of an 1-bit inverter.

The parameters can be extracted from any design system but they are highly technology dependent. Here, we use two different sets of technology parameters. The set indicated as *Motorola technology* is based on Motorola's H4C CMOS sea-of-gate design series [NB93]. The second technology is based on the VENUS design system [HNS86, Sie88]. Table 2.1 lists cost and delay of the basic components under either technology and figure 2.1 lists the symbols of the components. Besides that, we also supply the two binary constants 0 and 1 free of charge. They are special gates with one output but zero inputs.

Note that the Motorola technology provides NAND and NOR gates at the same cost as AND and OR gates but at half the delay. That provides the basis for the technology specific design optimizations, which are addressed in the sections 3.9 and 6.5.

### RAMs and ROMs

Summing up the cost of all RAM cells yields a simple formula of the cost of a RAM. However, this formula ignores the cost of the address decoder and the data select logic. Under Motorola technology ([NB93] page 6.5), the cost of RAMs obey a slightly different formula:

$$C_{ram}(A, d) = \begin{cases} C_{RAMcell} \cdot A \cdot d & \text{, Venus} \\ C_{RAMcell} \cdot (A + 3) \cdot (d + \log \log d) & \text{, Motorola} \end{cases}$$

According to [NB93], the Motorola technology provides small on-chip RAMs with access time

$$D_{ram}(A, d) = \log d + A/4 \quad , \quad A \le 64.$$

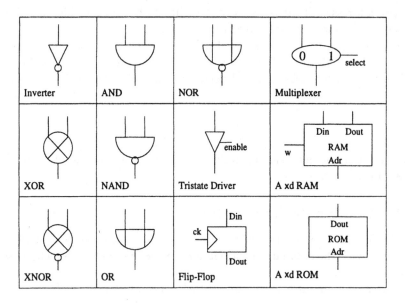

Figure 2.1: Symbols of the basic components

For all other RAMs, we apply the more general run time formula of Formella [For92]

$$D_{ram}(A, d) = R_1 \cdot \log A + R_2,$$

where $R_1$ models the delay of one of the $\log A$ levels of the address decoder and of the select logic. The term $R_2$ models the time required to write a single RAM cell respectively to read out the cell and to forward the data to the output of the RAM. For the Venus system, Formella derived the values $R_1 = 1.5$ and $R_2 = 0$ gate delays. Thus, the whole delay of this RAM is caused by its select and decode logic.

In our model, ROMs have the same delay as RAMs but they are usually cheaper by a factor $ROMfac$. The reference guide of the Motorola system [NB93] does not handle ROMs. Thus, we could assume a scaling factor of $ROMfac = 1$. However, that is not very realistic, because the VENUS system provides ROMs at one sixth of the cost of an equally sized RAM, and even scaling factors $ROMfac$ of 8 to 15 are possible [Yam90, KII90]. For the Motorola system, we therefore model the cost of a ROM by one eighth of the cost of an equally sized RAM.

## Multiport RAMs

Designers use multiport RAMs for efficient implementations of register files, because those structures often have to perform several reads and/or writes at the same time. For the DLX design of section 6, we need a dual-read single-write register file. This could be implemented by two RAMs with identical contents. They would write the same data at the same address in parallel or would read from two addresses. That is one way to realize such a register file, but cost and delay would be far from being realistic.

We therefore introduce a dual-read single-write RAM as a special component. The data in the Motorola Reference Guide ([NB93] pages 7.183–7.191) suggest the following estimate for cost and delay of such a component:

$$
\begin{aligned}
C_{ram2}(A, d) &= 1.6 \cdot C_{ram}(A, d) \\
D_{ram2}(A, d) &= 1.5 \cdot D_{ram}(A, d)
\end{aligned}
$$

The efficient implementation of multiport RAMs is a problem of VLSI design [WE85] rather than of circuit theory.

## 2.4.2   Combinatorial Circuits

Combining gates $G$ according to the rules of circuit theory [Weg87, Hot74] yields a circuit $S$ with the standard functionality. So, we only have to specify cost and delay of a circuit $S$.

The cost of a circuit S is the sum of the cost of all gates $G$ in S

$$
C_S = \sum_{G \in S} C_G.
$$

The delay of a circuit correlates to its depth, but in our model the delay (depth) of a gate depends on its type. The delay of a path $P$ from the inputs to an output of the circuit is the sum over all delays of the gates $G$ which lie on this path. The delay of the whole circuit $S$ is then the maximal delay of any path through the circuit.

$$
\begin{aligned}
D_P &= \sum_{G \text{ gate on path } P} D_G \\
D_S &= \max\{D_P \mid P \text{ path through S}\}
\end{aligned}
$$

**Example: n-bit Decoder**   The circuit of an n-bit decoder can recursively be defined, as illustrated in figure 2.2. The following formulae then describe its cost and delay:

$$
\begin{aligned}
C_{decs}(1) &= C_{inv} \\
C_{decs}(n) &= C_{decs}(n-1) + 2^n C_{and} + C_{inv} \\
\\
D_{decs}(1) &= D_{inv} \\
D_{decs}(n) &= \max\{D_{decs}(n-1), D_{inv}\} + D_{and}
\end{aligned}
$$

It is a messy, time consuming and error prone task to evaluate all formulae for cost and delay by hand. Formulae for delay are particularly bad, because they involve taking maxima and can generally not be simplified. We therefore

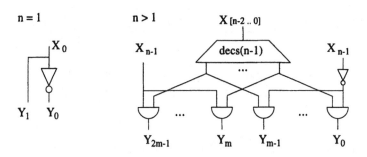

Figure 2.2: Recursive definition of the n-bit decoder circuitry; $m = 2^{n-1}$

write the formulae as C-routines. A computer can then keep track of cost and delay of the architecture. The following two routines define the cost of the n-bit decoder introduced in figure 2.2:

```
int cdecs(n)  int n;     /* Cost of n-bit decoder */
{return (n == 1 ? cnot : cdecs(n-1) + power2(n)*cand + cnot );}

int ddecs(n)  int n;
{return (n == 1 ? dnot : max(ddecs(n-1), dnot) + dand) ;}
```

Appendix B contains a complete list of all such C-routines used in this book.

### 2.4.3  Clocked Circuits

A register $R$ is a set of flipflops with a common clock input. A clocked circuit $S_c$ consists of a circuit $S$ and a set $R = \{R_1, \ldots, R_s\}$ of registers. They are arranged such that for all $i$, the clock signal $c_i$ and the input $Z_i$ of register $R_i$ is connected to outputs of combinatorial circuit $S$ and the output of register $R_i$ is fed back to the inputs of circuit $S$. This arrangement is drawn in figure 2.3.

The clocked circuit works in cycles $t = 1, 2, \ldots$. By induction over $t$ we define the values $R_i(t)$, $ce_i(t)$ and $Z_i(t)$. At the start of the computation all registers have a random but boolean content; this is the convention for powerup. Now suppose that after cycle $t$ each register $R_i$ has a well defined content $R_i(t)$. With these inputs the combinatorial circuit $S$ produces well defined outputs $c_i(t)$ and $Z_i(t)$ (because $S$ is cycle free). Then, we define for all $i$

$$R_i(t+1) = \begin{cases} Z_i(t) & \text{if} \quad ce_i(t) = 1 \\ R_i(t) & \text{if} \quad ce_i(t) = 0. \end{cases}$$

This is obviously well defined. It means that at the end of cycle $t+1$ exactly the registers with active clock signal $ce_i(t) = 1$ change their internal state to the value $Z_i(t)$ applied on their data input. We use the notation $ce_i$ rather than $c_i$ because we treat these signals more like clock enable signals than like clock signals.

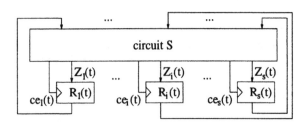

Figure 2.3: Schematics of a clocked circuit $S_c$ with combinatorial circuit $S$ and $s$ registers

Suppose $S_c$ is a clocked circuit with combinatorial circuit $S$ and registers $R_1, \ldots, R_s$ and for all $i$ register $R_i$ consists of $n_i$ flipflops. Then, the cost of $S_c$ is the cost of the combinatorial circuit $S$ plus the cost of the flipflops of the registers:

$$C_{Sc} = C_S + \sum_{i=1}^{s} n_i \cdot C_{ff}.$$

The cycle time $T_{Sc}$ of the clocked circuit equals the delay of the circuit $S$ plus the delay of the registers plus some time $\delta$ for setup times. To keep the formulae simple, $\Delta := D_{ff} + \delta$ expresses the time overhead for clocking a register. Thus,

$$T_{Sc} = D_S + D_{ff} + \delta = D_S + \Delta.$$

Here we use $\delta = 1$.

## 2.4.4 Modified Hardware

It remains to deal with RAMs, ROMs and tristate drivers. Dealing with ROM is easy. It behaves like a combinatorial circuit, where the delay equals the access time of the ROM. RAMs and drivers do not behave like combinatorial circuits or clocked circuits. However, if we fix the enable signal of a driver, then the driver behaves like a circuit, and if we fix the write signal of a RAM, then the RAM behaves like a combination of registers and a circuit.

### Modeling Tristate Drivers

A driver $d$ with an active enable signal $en = 1$ behaves like a circuit implementing the identity function. It forwards the input data signals to its output. With inactive enable signal $en = 0$, the driver behaves like there would be no connection between the data inputs and the outputs. For fixed control signals, the driver can therefore be modeled by the circuits $d(0)$ respectively $d(1)$ of figure 2.4. For $\xi \in \{0, 1\}$ we call the circuit $d(\xi)$ the *modified hardware* for driver $d$ and control signal $\xi$.

Figure 2.4: Modified hardware of the tristate driver $d$

## Modeling RAMs

The speed of a RAM is characterized by its access time $D_{ram}$. A RAM $r$ can perform two operations, read and write. The write signal $w$ indicates which operation should be performed. With an inactive write signal ($w=0$), data are read from the RAM. The RAM behaves like the combination $r(0)$ of registers and combinatorial circuit $S_r$ in figure 2.5, where the delay $D_{S_r}$ of circuit $S_r$ equals the access time $D_{ram}$ of the RAM. With an active write signal ($w=1$) data are written into the RAM. The RAM behaves like the combination $r(1)$ of registers and combinatorial circuit $S_w$ in figure 2.5, where

$$D_{S_w} + D_{ff} = D_{ram}.$$

For $\xi \in \{0,1\}$, we call the circuit $r(\xi)$ the *modified hardware* for RAM $r$ and control signal $\xi$.

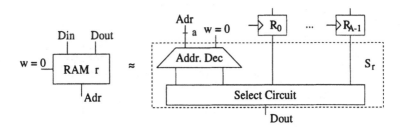

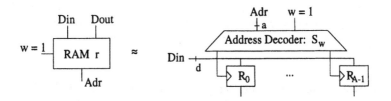

Figure 2.5: Modified hardware of the $A \times d$ RAM $r$; $a = \log A$

In the next subsection we will use the concept of modified hardware in order to define the functionality of control driven circuits. In this process combinatorial circuits, which have no open inputs in the original hardware, can get open inputs in the modified hardware. Thus, devices can get an undefined input signal $\omega$. There are three situations, where a device with an open input $\omega$ can produce an output in $\{0, 1\}$, namely:

1. $0 \text{ AND } \omega = \omega \text{ AND } 0 = 0, \quad 0 \text{ NAND } \omega = \omega \text{ NAND } 0 = 1$

2. $1 \text{ OR } \omega = \omega \text{ OR } 1 = 1, \quad 1 \text{ NOR } \omega = \omega \text{ NOR } 1 = 0$

3. a multiplexer with one undefined data input that is not selected forwards the other data input.

## 2.4.5 Control Driven Circuits

Control driven circuits $S_{cd}$ comprise two parts, namely the data paths DP and the control CON (figure 2.6). They are connected by three busses: bus $C_{out}$ from CON to DP and busses $N_{in}, C_{in}$ from DP to CON. In the data paths all kinds of components can occur, and they can be interconnected in very complicated ways. At this point we only demand that all clock signals, write signals, and enable signals of components in DP directly come from the control via the bus $C_{out}$. There may be additional signals on the bus $C_{out}$ which go into gates of the data paths, e.g., into function select inputs of an ALU. Restrictions about how to wire up the data paths will follow shortly from a very simple and general condition.

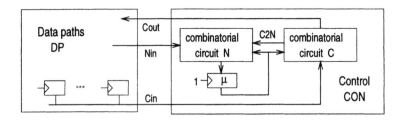

Figure 2.6: Block diagram of a control driven circuit with data paths DP and control CON. Signals on the bus $C_{in}$ are combinatorial, i.e., they come directly from registers in DP.

The control consists of a register $\mu$ and two combinatorial circuits $C$ and $N$. The content of register $\mu$ is called the *micro state*. The clock enable signal of register $\mu$ is tied to 1, i.e., this register is updated every cycle. Circuit $C$ computes the control signals of the data paths and puts them on the bus $C_{out}$. It also forwards some signals to circuit $N$ via bus $C2N$. The inputs of circuit $C$ are connected to the outputs of register $\mu$ and to the bus $C_{in}$ with signals from the data paths. We require that signals on this bus come directly from registers in the data paths. Circuit $N$ computes the next state of the control. Its inputs are connected to the outputs of register $\mu$, to bus $C2N$ and to bus $N_{in}$ with signals from the data paths.

### Functionality

A control driven circuit $S_{cd}$ works in cycles $t = 1, 2, \ldots$. Initially, all registers have fixed boolean content. Suppose all registers and RAMs have a well defined content after round $t$. Then all inputs of circuit $C$ are well defined, because they come directly from registers. Hence, the current outputs $C_{out}$ of this circuit are well defined. By the rules of the previous subsection, this output $C_{out}$ defines a modified data paths $DP(C_{out})$. Output $C_{out}$ is called *admissible* if the following conditions hold:

1. The modified data paths $DP(C_{out})$ together with the control $CON$ is a clocked circuit, which we denote by $S_{cd}(C_{out})$. The underlying combinatorial circuit may have open inputs.

2. In circuit $S_{cd}(C_{out})$, all data inputs of registers with an active clock enable signal have values in $\{0, 1\}$.

Registers and RAMs are now updated according to the rules for clocked circuits.

### Cost and Cycle Time

The cost of a control driven circuit $S_{cd}$ is the sum of the cost of all its components. The cycle time of $S_{cd}$ is the maximum of the cycle time of all clocked circuits $S_{cd}(\xi)$; $\xi$ ranges over all values the bus $C_{out}$ can assume by the rules of the previous paragraph:

$$T_{S_{cd}} = \max\{T_{S_{cd}(\xi)} \mid \xi \text{ admissible value of } C_{out}\}.$$

## 2.4.6  A Special Device: Main Memory

We do not study the cost or realization of main memory here. Nevertheless, we have to specify the functionality of main memory in such a way, that it can act like a device in a control driven circuit. In the simplest case, we could specify main memory as a big RAM which has a certain access time $d_{mem}$ and no cost. Moreover, at the start of the first cycle (i.e., after powerup), main memory contains the initial program starting at address 0 as well as the data for that program.

For the design of the DLX machine in sections 6 and 8 this does not suffice because main memory consists of several banks and provides some status signals.

**Multiple Memory Banks**  Main memory consists of 4 banks, where each bank is 1 byte wide. Reads always affect all banks. Writes are controlled by 4 write signals $w_0, \ldots, w_3$, one for each bank.

**Memory Status Signals**  Main memory provides two output flags for status information. The *busy* flag indicates that main memory can not complete the access in the current cycle of the hardware. The flag *pagefault* signals that the requested data lies on a memory block (page) which is not available during the current cycle.

While the flag *busy* is active during a read access, main memory provides a fixed, well defined boolean value on its data output. However, this value is not likely to be the requested memory data. Technically speaking, this could be implemented by a pullup resistor on the data output of main memory.

Main memory may activate one or both status signals (*pagefault* = 1 or *busy* = 1) during the current cycle of the hardware. The status signal is inactive during the whole cycle, if main memory does not activate the signal in the same cycle.

**Temporal Behavior**  We use the parameters $d_{mem}$ and $d_{mstat}$ to model the time behavior of main memory. The parameter $d_{mem}$ defines the (minimal) data access time of main memory. Main memory requires the memory status time $d_{mstat}$ to generate its status signals.

One would like the memory status time to be much shorter than the access time $d_{mem}$. If not stated otherwise, we assume a memory status time of $d_{mstat} = 5$ gate delays. Note, that even for an on-chip cache, that is an overly optimistic assumption. Most of the designs presented in this book tolerate a memory status time of 20 to 25 gate delays, but for some optimized designs of section 7 the memory status time lies on the critical path. Those designs would call for a two cycle memory access.

In this framework, we would like to avoid those additional memory cycles in order to keep the designs simple and to be in line with designs like MIPS R10000 [Gwe94], Pentium [Mes94] and AMD K5 [Sla94]. These processors manage with one cycle cache accesses even for large on-chip caches. Thus, we choose a memory status time $d_{mstat} = 5$ which is small enough to avoid those conflicts. However, in section 6.5 (exercise 6.4), we will discuss an optimization of the control unit which allows us to tolerate a memory status time of roughly 65 to 84% of the DLX cycle time.

**Functionality**  A memory bank $b_i$ with an active write signal ($w_i = 1$) performs a write. Under an inactive write signal ($w_i = 0$), the bank performs a read. Therefore, a memory bank basically behaves like a RAM which is of the same size and uses the write signal $w_i$. For a control signal $w_i \in \{1, 0\}$, we denote the modified hardware of bank $b_i$ by $b_i(w_i)$.

Under the control signals $w = (w_0, \ldots, w_3) \in \{0, 1\}^4$, the functionality of main memory $m$ is modeled by its *modified hardware* $m(w)$. The modified hardware of main memory behaves like the combination of the modified hardware $b_i(w_i)$ of the four memory banks $b_i$ and the combinatorial circuit $S_{mstat}$ in figure 2.7. This circuit generates the status signals *busy* and *pagefault*.

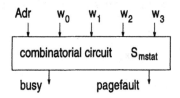

Figure 2.7: Circuit $S_{mstat}$ of the modified hardware of main memory

### 2.4.7   Powerup Convention and External Signals

After *powerup*, i.e., at the start of the first cycle, all registers and RAMs have a random but boolean content. ROMs contain their specific data, and main memory contains the initial program starting at address 0 and the data for that program.

The designer has to guarantee that under those conditions the machine starts its computation properly at the first instruction of the initial program. That calls for an external signal which indicates powerup.

So far, our model has no device that generates external signals. This is easily fixed: an *external input* is simply a signal that can be used like the output of a flipflop and that may change its value at the beginning of every cycle. These signal toggle at most once per cycle. Besides powerup, external inputs are also used to initiate a hardware reset or for signals from external I/O devices (section 8).

### 2.4.8   Extended Nomenclature of Delay Formulae

For the sake of clarity, designs are specified in a modular fashion, i.e., they are composed of several sub-circuits. That also makes it easier to manage design changes. Since our analysis shall reflect this modularity as well, we first compute the delay of the sub-circuits, and in a second step, we combine these delays to the cycle time of the whole hardware. In this context, a run time estimate requires to consider the delay of certain paths through a sub-circuit. In order to easily express these delays, we now extend the formalism of the delay formulae.

**Delay of Circuits**

Let $S$ be a combinatorial circuit with inputs $I$ and outputs $O$. By $D_S(d_i \, ; \, r_j)$, we denote the maximum of the delays of all paths $P$ through $S$ from input $d_i \in I$ to output $r_j \in O$, if such a path does exist:

$$
D_S(d_i \, ; \, r_j) = \begin{cases} \max\left\{ D_P \,\middle|\, \begin{array}{l} \text{P path through } S \\ \text{from } d_i \text{ to } r_j \end{array} \right\} & ; \text{ if P exists} \\ \qquad\qquad 0 & ; \text{ otherwise} \end{cases}
$$

Let $I' \subset I$ be a subset of the inputs and $O' \subset O$ a subset of the outputs of circuit $S$. By

$$D_S(I'; O') = \max\{D_S(d_i; r_j) \mid d_i \in I', r_j \in O'\},$$

we denote the delay of circuit $S$ from inputs $I'$ to outputs $O'$, i.e., the maximum of the delays of all paths through $S$ from an input in $I'$ to an output in $O'$. For $I' = I$, the set $I'$ does not restrict the paths through circuit $S$, and thus we omit it in the delay formula of $S$. The same holds for $O' = O$. The maximal delay of the combinatorial circuit $S$ can therefore be expressed as

$$D_S(I; O) = D_S.$$

Omitting one of the sets $I$ and $O$ can cause the ambiguity whether the remaining set of signals correspond to inputs or to outputs of the circuit. In such a case, we omit the set but keep the semicolon. Thus, the following three formulae express the same delay:

$$D_S(I; O') = D_S(; O') = D_S(O').$$

Some circuits are closer specified by a list of parameters, like the n-bit decoder $decs(n)$ or the $A \times n$ RAM. For those circuits, we treat the list of parameters as an additional argument of the delay formula. Thus, the delay of an n-bit decoder is expressed as $D_{decs}(I; O; n) = D_{decs}(n)$.

### Accumulated Delay

For a clocked circuit, comprising several sub-circuits, the delay formula shall express whether the underlying paths do start in a register or not. For that purpose, we introduce the *accumulated delay* of a circuit.

Let the combinatorial circuit $S$ with inputs $I$ and outputs $O$ be part of a clocked circuit $S_c$. For $I' \subset I$ and $O' \subset O$, we denote by the accumulated delay $A_S(I'; O')$ the maximum of the delays of all paths through circuit $S$ to one of its outputs $r_j \in O'$; the paths must start in a register of $S_c$ and must enter circuit $S$ via an input $d_i \in I'$. If all inputs in $I'$ of circuit $S$ are connected directly to registers, then

$$A_S(I'; O') = D_S(I'; O').$$

This formalism can also be applied to the cycle time of the clocked circuit $S_c$. $T_S(I'; O')$ then specifies the maximum of the cycle times of all cycles in $S_c$ which enter circuit $S$ via an input $d_i \in I'$ and leave it via an output $r_j \in O'$. This is well defined, even if neither the inputs $I'$ nor the outputs $O'$ are directly connected to registers.

The following example shall illustrate, how this formalism can be used to express the cycle time of a clocked circuit.

**Example:** The clocked circuit $S_c$ of figure 2.8 comprises three cycles:

- leaving circuit $S_1$ via output $d_3$,

- entering circuit $S_2$ via input $d_1$,

- entering circuit $S_2$ via input $d_2$.

Thus, the cycle time of circuit $S_c$ can be expressed as

$$T_{S_c} \;=\; \max\{T_{S_1}(;d_3), T_{S_2}(d_1;), T_{S_2}(d_2;)\} \;=\; \max\{T_{S_1}(d_3), T_{S_2}(d_1), T_{S_2}(d_2)\},$$

with

$$
\begin{array}{rcl}
T_{S_1}(d_3) &=& A_{S_1}(d_3) + \Delta \;=\; D_{S_1}(d_3) + D_{ff} + \delta \\
T_{S_2}(d_1) &=& A_{S_2}(d_1) + \Delta \;=\; D_{S_2}(d_1) + D_{ff} + \delta \\
T_{S_2}(d_2) &=& A_{S_2}(d_2) + \Delta \;=\; A_{S_1}(d_2) + D_{S_2}(d_2) + D_{ff} + \delta.
\end{array}
$$

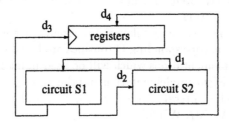

Figure 2.8: Schematic of a clocked circuit $S_c$

# Chapter 3

# Functional Modules

Before focusing on the design of the DLX machine, we introduce some important circuits and emphasize the subtleties of the design of an arithmetic unit for two's complement numbers (theorems 3.2 and 3.3). Readers familiar with this material may glance over this section.

For simplicity's sake, cost and delay of the circuits are expressed by mathematical equations. The corresponding C-routines are listed in appendix B.3. We only describe the implementations which will be used in later sections. A formal definition, further efficient implementation and the proof of their correctness can be found in the standard literature on circuit theory and computer architecture like: [Spa76, Sav87, Weg87, WH90].

## 3.1  N-bit Components

Most of the basic components introduced in section 2.4.1 deal with single bits, but designs also use wider data paths. For describing the hardware, we therefore need circuits which basically behave like the basic components but can handle n-bit wide binary data.

**Definition 3.1 (n-bit Multiplexer)** *An n-bit multiplexer is a circuit which computes the function*   $mux_n : \{0, 1\}^{2n+1} \rightarrow \{0, 1\}^n$,

$$mux_n(a_{n-1}, \ldots, a_0, b_{n-1}, \ldots, b_0, s) = \begin{cases} a & for \quad s = 0 \\ b & for \quad s = 1. \end{cases}$$

The n-bit multiplexer ($mux_n$) is implemented by $n$ multiplexers. These $n$ basic components are all controlled via a common select signal, as figure 3.1 indicates. The circuit has the same delay as the basic component but with $n$ times the cost. Consequently, the basic multiplexer is an 1-bit multiplexer. If there is no possibility of confusion, we omit the width $n$.

$$C_{mux}(n) = n \cdot C_{mux}$$
$$D_{mux}(n) = D_{mux}$$

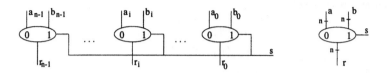

Figure 3.1: Circuit and symbol of an n-bit multiplexer

Note, especially for large $n$, this cost and run time assumption is too optimistic because a single select signal has to drive a lot of gates. The select input of the n-bit multiplexer is called to have a high *fanout*. High fanout increases the delay of the multiplexer. Standard design systems therefore provide additional hardware to amplify the control signal. That lowers the run time impact of the fanout, but cost and delay are still underestimated. The same holds for other n-bit components with single control signal, like n-bit tristate drivers and n-bit registers.

However, our model ignores the impact of fanout. Nevertheless, we express the delay of the n-bit multiplexer (and similar components) as a function of $n$; that makes it easier to include fanout considerations in a later version of the model (see exercises of section 3.9).

**n-bit Register**  An n-bit register $ff_n$ is a circuit consisting of $n$ flipflops with input $a = (a_{n-1}, \dots, a_0) \in \{0, 1\}^n$, output $b = (b_{n-1}, \dots, b_0) \in \{0, 1\}^n$, and common clock signal $ce \in \{0, 1\}$ (figure 3.2a). The n-bit register has the same delay as a flipflop but it is $n$ times more expensive.

$$
\begin{aligned}
C_{ff}(n) &= n \cdot C_{ff} \\
D_{ff}(n) &= D_{ff}
\end{aligned}
$$

**n-bit Tristate Driver**  An n-bit tristate driver $driv_n$ consists of $n$ 1-bit tristate drivers with input $a = (a_{n-1}, \dots, a_0) \in \{0, 1\}^n$, output $b = (b_{n-1}, \dots, b_0) \in \{0, 1\}^n$, and common enable signal $oe \in \{0, 1\}$ (figure 3.2b). The n-bit driver has the same delay as the basic component but it is $n$ times more expensive.

$$
\begin{aligned}
C_{driv}(n) &= n \cdot C_{driv} \\
D_{driv}(n) &= D_{driv}
\end{aligned}
$$

**n-bit Wide Gates**  An n-bit inverter circuit computes the function

$$
inv_n : \{0, 1\}^n \rightarrow \{0, 1\}^n; (a_{n-1}, \dots, a_0) \mapsto (\overline{a_{n-1}}, \dots, \overline{a_0}).
$$

This circuit is implemented as n 1-bit inverters which work in parallel (figure 3.2c). The n-bit inverter therefore has the same delay as the basic component but at n times the cost.

$$C_{inv}(n) = n \cdot C_{inv}$$
$$D_{inv}(n) = D_{inv}$$

Let $\bullet$ be any of the dyadic functions AND, NAND, OR, NOR, XOR and XNOR. The n-bit circuit of $\bullet$ computes the function

$$\bullet_n : \{0, 1\}^{2n} \rightarrow \{0, 1\}^n;$$

$$(a_{n-1}, \ldots, a_0, b_{n-1}, \ldots, b_0) \mapsto (a_{n-1} \bullet b_{n-1}, \ldots, a_0 \bullet b_0).$$

Let $G_\bullet$ be the gate implementing the dyadic function $\bullet$. The n-bit circuit of $\bullet$ then consists of n gates $G_\bullet$ with inputs $a_i$, $b_i$ and output $r_i$ (figure 3.2d). The n-bit version has the same delay as gate $G_\bullet$ but at n times the cost.

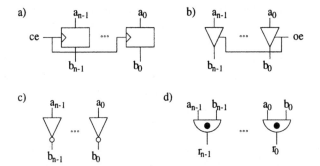

Figure 3.2: Circuits of an n-bit register (a), of an n-bit tristate driver (b), of an n-bit inverter (c) and of an n-bit dyadic function $\bullet$ (d)

For example, cost and delay of an n-bit AND gate run at

$$C_{and}(n) = n \cdot C_{and}$$
$$D_{and}(n) = D_{and}.$$

## 3.2  Decoders

**Definition 3.2** *For $x \in \{0, 1\}^n$, we denote by*

$$<x> = <x_{n-1}, \ldots, x_0> = \sum_{i=0}^{n-1} x_i 2^i$$

*the number represented by x. We often refer to x as n-bit binary number. For $y \in \{0, \ldots, 2^{n-1}\}$ with $y = <x>$, $x \in \{0, 1\}^n$, we denote by*

$$bin_n(y) = (x_{n-1}, \ldots, x_0) = x$$

*the n-bit binary representation of y.*

**Definition 3.3 (n-bit Decoder)** *An n-bit decoder maps an n-bit input into a* $2^n$*-bit output. For any input* $x \in \{0, 1\}^n$*, the* $<x>$*-th bit of the output is active (1), all others are 0.*

Figure 3.3 depicts the recursive definition of a simple n-bit decoder circuit. We already introduced this circuit in section 2.4.2. This implementation has the

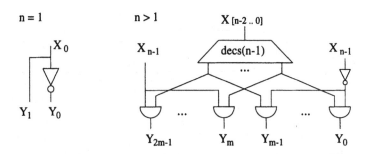

Figure 3.3: Recursive definition of an n-bit decoder. To simplify the indices, we use the abbreviation: $m = 2^{n-1}$.

following cost and delay:

$$
\begin{aligned}
C_{decs}(1) &= C_{inv} \\
C_{decs}(n) &= C_{decs}(n-1) + 2^n\, C_{and} + C_{inv} \\
D_{decs}(1) &= D_{inv} \\
D_{decs}(n) &= \max\{D_{decs}(n-1), D_{inv}\} + D_{and}
\end{aligned}
$$

A detailed analysis of the circuit in figure 3.3 indicates, that for each output bit, $n$ literals are combined via AND gates. Arranging these gates in a balanced binary tree reduces the delay of the decoder from order $n$ to order $\log n$. That is the concept behind our second implementation specified in figure 3.4. The 1-bit decoder remains the same. For larger $n$, the decoder is composed of an $\lceil n/2 \rceil$-bit decoder, an $\lfloor n/2 \rfloor$-bit decoder and $2^n$ and gates. Its cost and delay therefore run at:

$$
\begin{aligned}
C_{decf}(1) &= C_{inv} \\
C_{decf}(n) &= C_{decf}(\lceil n/2 \rceil) + C_{decf}(\lfloor n/2 \rfloor) + 2^n \cdot C_{and} \\
D_{decf}(1) &= D_{inv} \\
D_{decf}(n) &= D_{decf}(\lceil n/2 \rceil) + D_{and}
\end{aligned}
$$

Figures 3.5 and 3.6 list the cost and the delay of both decoder designs. The second implementation results in faster and cheaper decoders. Large decoders then have only half the cost, and the delay grows much more slowly than for the first implementation. From now on, we therefore only use the second decoder design. We denote its cost and delay by $C_{dec}$ and $D_{dec}$.

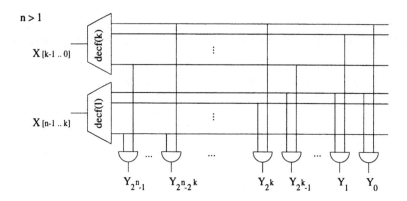

Figure 3.4: Implementation of a fast n-bit decoder. To simplify indices, we use $k = \lceil n/2 \rceil$ and $l = \lfloor n/2 \rfloor$.

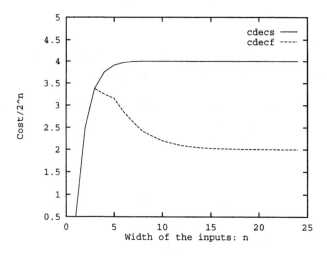

Figure 3.5: Cost comparison of the two decoder designs. The cost are normalized by the factor $2^n$

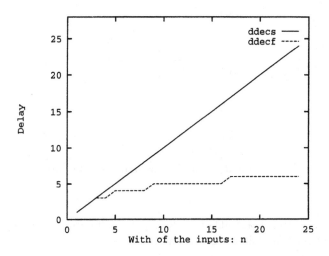

Figure 3.6: Run time comparison of the two decoder designs

## 3.3   A Half-Decoder

**Definition 3.4 (n-bit Half-Decoder)** *An n-bit half-decoder maps an n-bit input into a $2^n$-bit output. If the input has value $x \in \{0, 1\}^n$, then the $<x>$ least significant bits of the output are active (1), the remaining bits are 0.*

The circuit in figure 3.7 implements an n-bit half-decoder at the cost $C_{hdec}(n)$ and the delay $D_{hdec}(n)$. This circuit is a modification of the first decoder implementation.

$$
\begin{aligned}
C_{hdec}(1) &= 0 \\
C_{hdec}(n) &= C_{hdec}(n-1) + 2^{n-1} \cdot (C_{and} + C_{or}) \\
D_{hdec}(1) &= 0 \\
D_{hdec}(n) &= D_{hdec}(n-1) + \max\{D_{and}, D_{or}\}
\end{aligned}
$$

## 3.4   An Encoder

**Definition 3.5 (n-bit Encoder)** *An n-bit encoder maps a $2^n$-bit input into an n-bit output. Exactly one bit of the input has to be 1, all other bits are 0. If the j-th bit of the input is active (1), then the output is $bin_n(j)$.*

An encoder computes the inverse function of a decoder. We first consider a slightly different encoder with an additional output flag $F$. This flag tests, whether at least one bit of the input is active or not. If the input is 0, then the

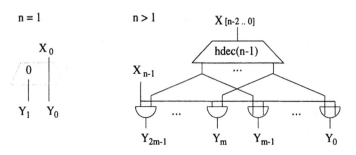

Figure 3.7: Recursive specification of an n-bit half-decoder. $m = 2^{n-1}$

flag and the standard output are also 0. The circuit of figure 3.8 implements this modified encoder at the following cost and delay:

$$
\begin{aligned}
C_{enf}(1) &= C_{or} \\
C_{enf}(n) &= 2 \cdot C_{enf}(n-1) + n \cdot C_{or} \\
D_{enf}(1) &= D_{or} \\
D_{enf}(n) &= D_{enf}(n-1) + D_{or}
\end{aligned}
$$

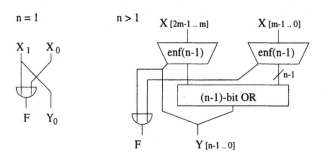

Figure 3.8: Recursive specification of an n-bit encoder with special output flag $F$, $m = 2^{n-1}$

Omitting the special output flag makes the circuit cheaper. The standard n-bit encoder can then be implemented as indicated in figure 3.9. This circuit has cost $C_{enc}(n)$ and delay $D_{enc}(n)$:

$$
\begin{aligned}
C_{enc}(1) &= 0 \\
C_{enc}(n) &= C_{enf}(n-1) + C_{enc}(n-1) + (n-1) \cdot C_{or} \\
D_{enc}(1) &= 0 \\
D_{enc}(n) &= \max\{D_{enf}(n-1),\, D_{enc}(n-1)\} + D_{or}
\end{aligned}
$$

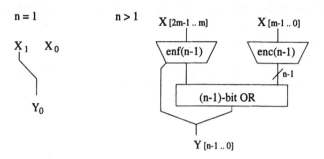

Figure 3.9: Recursive specification of a standard n-bit encoder, $m = 2^{n-1}$

## 3.5  Testing for Zero

**Definition 3.6 (n-bit Zero Test)** *The circuit of an n-bit zero test maps its n-bit input $x$ into a binary output $z$. The output has the value 1 iff all bits $x_i$ of the input are 0. Otherwise, the output bit has the value $z = 0$.*

The result $z$ of an n-bit zero test of the input $x$ can also be expressed as

$$z = \overline{x_0 \vee x_1 \vee \cdots \vee x_{n-1}}.$$

The n-bit zero test can therefore be implemented with $n - 1$ OR gates and one inverter. When arranging the OR gates in a balanced binary tree, the circuit has logarithmic delay. The result of the tree has to be passed through the inverter. In our implementation, we replace the last gate of the or-tree and the inverter by a NOR gate. The circuit then has the cost

$$C_{zero}(n) = (n - 2)\, C_{or} + C_{nor}$$

and the delay

$$D_{zero}(n) = (\lceil \log n \rceil - 1) \cdot D_{or} + D_{nor}.$$

## 3.6  The Ripple Carry Incrementer

**Definition 3.7 (n-bit Incrementer)** *An n-bit incrementer has an n-bit input and an $(n + 1)$-bit output. It computes the function $inc_n : \{0, 1\}^n \to \{0, 1\}^{n+1}$, $inc_n(x) = bin_{n+1}(<x> +1)$. The most significant bit of the result indicates whether an overflow occurred or not.*

There exist many implementations of incrementers. Since it is a special case of the addition — the second input is always 1 — any adder design can be adapted to an incrementer circuit. The ripple carry incrementer of figure 3.10

is quite a cheap implementation with the following cost and delay:

$$C_{Inc}(n) \;=\; (n-1)(C_{and}+C_{xor})+C_{inv}$$

$$D_{Inc}(n) \;=\; \begin{cases} D_{inv} & , n = 1 \\ \max\{D_{inv},\, (n-1)\cdot D_{and}, \\ \quad (n-2)\cdot D_{and}+D_{xor}\} & , n > 1 \end{cases}$$

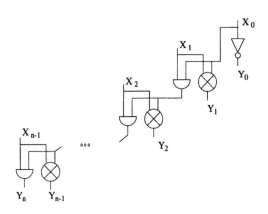

Figure 3.10: Circuit of an n-bit ripple carry incrementer $Inc_n$

## 3.7 Arithmetic Circuits

The core of each arithmetic logic unit (ALU) is a circuit that adds or subtracts two numbers. So far, we only represented positive numbers. In order to perform subtraction in all its generality, we also have to handle negative numbers. The *two's complement representation* is commonly used to represent both positive and negative numbers.

**Definition 3.8 (n-bit Two's Complement)** *For $x \in \{0, 1\}^n$, we denote by*

$$[x] \;=\; [x_{n-1},\ldots,x_0] \;=\; -x_{n-1}2^{n-1}+\sum_{i=0}^{n-2} x_i 2^i$$

*the value of the number y represented by x. The number y lies within the range $T_n := \{-2^{n-1},\ldots,2^{n-1}-1\}$. We denote by*

$$two_n(y) \;=\; (x_{n-1},\ldots,x_0) \;=\; x$$

*the n-bit two's complement representation of y. We often refer to x as n-bit two's complement number.*

The following characteristics of two's complement numbers are essential for the construction of the arithmetic unit, because they indicate that the addition and subtraction of two's complement numbers can be performed by an adder for binary numbers. The proofs of the two theorems can be found in [KP95].

**Theorem 3.1** *Let $x \in \{0, 1\}^n$ then $[(x_{n-1}, x_{n-1}, \ldots, x_0)] = [x]$ and*

$$-[x] \;=\; [\bar{x}] + 1 \;=\; [(\overline{x_{n-1}}, \ldots, \overline{x_0})] + 1.$$

**Theorem 3.2** *For $a, b \in \{0, 1\}^n$ and $cin \in \{0, 1\}$, let $s \in \{0, 1\}^n$ and $c_n \in \{0, 1\}$ be such that*

$$<c_n, s> \;=\; <a> + <b> + cin.$$

*For all $1 \le i \le n$, let bit $c_i$ indicate the carry from position $i - 1$ to position $i$, i.e.,*

$$<c_i, s_{i-1}, \ldots, s_0> \;=\; <a_{i-1}, \ldots a_0> + <b_{i-1}, \ldots b_0> + cin.$$

*Then the following two conditions hold:*

1. *The sum $z$ of the numbers $[a]$, $[b]$ and $cin$ lies in the set $T_n$ iff the two carries $c_n$ and $c_{n-1}$ are the same;*

$$z := [a] + [b] + cin \in T_n \quad \Longleftrightarrow \quad c_n = c_{n-1}$$

2. *If the sum $z$ lies in $T_n$ it can be expressed as $z = [(s_{n-1}, \ldots, s_0)]$.*

### 3.7.1  The Two's Complement Adder

**Definition 3.9 (Adder)** *An n-bit adder is a circuit which computes the function $+_n : \{0, 1\}^{2n+1} \to \{0, 1\}^{n+3}$,*

$$+_n(a_{n-1}, \ldots, a_0, b_{n-1}, \ldots, b_0, cin) = (neg, ovf, c_n, s_{n-1}, \ldots, s_0)$$

*with*

$$<c_n, s> \;=\; <a> + <b> + cin.$$

*The flag $ovf$ indicates an overflow, i.e., $z := [a] + [b] + cin \notin T_n$, and the flag $neg$ indicates that the sum $z$ is negative. The input flag $cin$ is called carry-in.*

By induction on $n$, it is easy to show that the following formulae express the relation between the inputs $a, b \in \{0, 1\}^n$ and $cin \in \{0, 1\}$ of an n-bit adder and its output bits $s \in \{0, 1\}^n$ and $c_n \in \{0, 1\}$:

$$\begin{aligned}
s_i &= a_i \otimes b_i \otimes c_i & 0 \le i < n \\
c_{i+1} &= (a_i \otimes b_i) \wedge c_i \vee (a_i \wedge b_i) \\
c_0 &= cin
\end{aligned} \tag{3.1}$$

## The Flag *neg*

For $a, b \in \{0, 1\}^n$ and $cin \in \{0, 1\}$, signal *neg* shall indicate that the sum $z = [a] + [b] + cin$ is negative.

Let $x, y \in T_n$ be such that $x = [a]$ and $y = [b]$. The sum $z = x + y + cin$ lies in $T_{n+1}$ but not necessary in $T_n$, we therefore have to use $(n+1)$-bit arithmetic. According to theorem 3.1, the sum $z$ can then be written as

$$z = [a] + [b] + cin = [(a_{n-1}, a)] + [(b_{n-1}, b)] + cin.$$

Let $s \in \{0, 1\}^{n+1}$ be such that $s = two_{n+1}(z)$. The sum bit $s_n$ is then defined as

$$s_n = a_n \otimes b_n \otimes c_n = a_{n-1} \otimes b_{n-1} \otimes c_n.$$

Since a two's complement number $s \in \{0, 1\}^{n+1}$ is negative iff $s_n = 1$, the flag *neg* defined as

$$neg = c_n \otimes a_{n-1} \otimes b_{n-1}.$$

really indicates that the sum $z = [a] + [b] + cin$ of $a, b \in \{0, 1\}^n$ and $cin \in \{0, 1\}$ is negative.

## The Overflow Flag *ovf*

For inputs $a, b \in \{0, 1\}^n$ and $cin \in \{0, 1\}$, signal *ovf* shall indicate an overflow, i.e., the sum $z = [a] + [b] + cin$ does not lie in $T_n$. According to theorem 3.2, that is the case iff the two carries $c_n$ and $c_{n-1}$ have different values. Thus, the flag *ovf* can simply be expressed as

$$ovf = c_n \otimes c_{n-1}.$$

However, this test is not feasible for adder implementations which do not explicitly generate these carry bits. The following theorem [Spa76, Hot72] shows how under those conditions $(n+1)$-bit arithmetic can be used in order to detect the overflow.

**Theorem 3.3** *For $a, b \in \{0, 1\}^n$, $cin \in \{0, 1\}$, $a_n = a_{n-1}$, and $b_n = b_{n-1}$, let $s \in \{0, 1\}^n$ and $c_n, c_{n+1}, s_n \in \{0, 1\}$ be such that*

$$\langle c_n, s \rangle = \langle a \rangle + \langle b \rangle + cin,$$
$$\langle c_{n+1}, s_n, s \rangle = \langle a_n, a \rangle + \langle b_n, b \rangle + cin,$$

*then:*

1. *the sum $z$ of the numbers $[a]$, $[b]$ and $cin$ lies in the set $T_n$ iff $s_n$ equals $s_{n-1}$,*

$$z := [a] + [b] + cin \in T_n \iff s_n = s_{n-1}$$

2. *the overflow can be detected by*

$$ovf = s_{n-1} \otimes s_n = s_{n-1} \otimes a_{n-1} \otimes b_{n-1} \otimes c_n.$$

**Proof:**

According to formula 3.1, the sum bits $s_{n-1}$ and $s_n$ are defined as

$$
\begin{aligned}
s_{n-1} &= a_{n-1} \otimes b_{n-1} \otimes c_{n-1} \\
s_n &= a_n \otimes b_n \otimes c_n = a_{n-1} \otimes b_{n-1} \otimes c_n.
\end{aligned}
$$

Thus, $s_{n-1}$ equals $s_n$ iff $c_{n-1}$ equals $c_n$,

$$
s_{n-1} = s_n \iff c_{n-1} = c_n.
$$

According to theorem 3.2, that is the condition for the sum $z$ to be in $T_n$. As a consequence, an overflow ($z \notin T_n$) can be detected by

$$
ovf = s_{n-1} \otimes s_n = s_{n-1} \otimes a_{n-1} \otimes b_{n-1} \otimes c_n.
$$

$\square$

Many implementations of the n-bit adder are known [Spa76, Weg87]. They have different cost and delay. In the following, we present three implementations know as ripple carry adder, carry look-ahead adder and conditional sum adder.

## 3.7.2  The Ripple Carry Adder (RCA)

In the previous section, we saw that the sum $(c_n, s_{n-1}, \ldots, s_0)$ of two numbers $a, b \in \{0, 1\}^n$ and input carry $c_0 \in \{0, 1\}$ can be expressed as:

$$
\begin{aligned}
s_j &= a_j \otimes b_j \otimes c_j & , 0 \le j < n \\
c_{j+1} &= (a_j \otimes b_j) \wedge c_j \vee (a_j \wedge b_j) & , 0 \le j < n.
\end{aligned} \tag{3.2}
$$

A *full adder FA* is a circuit which computes the 1-bit addition

$$
add_1 : \{0, 1\}^3 \to \{0, 1\}^2; (a, b, c_i) \mapsto (c_o, s).
$$

Implementing the full adder along the lines of the recursion formula shown above, yields the circuit of figure 3.11 with cost

$$
C_{FA} = 2 \cdot C_{xor} + 2 \cdot C_{and} + C_{or}.
$$

Since the circuit is symmetrical in its inputs $a$ and $b$, the following delay formulae can be derived:

$$
\begin{aligned}
D_{FA}(a, b; s) &= 2 \cdot D_{xor} \\
D_{FA}(a, b, c_i; c_o) &= D_{xor} + D_{and} + D_{or} \\
D_{FA}(c_i; c_o) &= D_{and} + D_{or} \\
D_{FA}(c_i; c_o, s) &= \max\{D_{xor}, D_{and} + D_{or}\}.
\end{aligned}
$$

Formula 3.2 computes recursively the sum and the carry bits. The output carry $c_o$ of level $j$ serves as input carry $c_i$ of level $j + 1$. Thus, the n-bit *ripple*

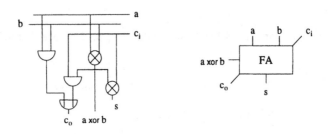

Figure 3.11: Circuit of a full adder FA

*carry adder* $RCA_n$ cascades $n$ full adders as indicated in figure 3.12. That takes care of the sum and carry bits. The flags *neg* and *ovf* are generated along the lines of theorems 3.1 and 3.2

$$neg = a_{n-1} \otimes b_{n-1} \otimes c_n$$
$$ovf = c_n \otimes c_{n-1}.$$

The full adder responsible for the sum bit $s_{n-1}$ also provides $a_{n-1} \otimes b_{n-1}$. Thus, the n-bit ripple carry adder can be implemented at the following cost and delay:

$$C_{RCA}(n) = n \cdot C_{FA} + 2 \cdot C_{xor}$$

$$D_{RCA}(n) = D_{xor} +$$
$$\begin{cases} \max\{D_{FA}(a,b,c_i;c_o), D_{FA}(a,b;s), D_{FA}(c_i;c_os)\} & , n = 1 \\ \max\{D_{FA}(a,b;s), D_{FA}(a,b,c_i;c_o) \\ \quad +(n-2) \cdot D_{FA}(c_i;c_o) + D_{FA}(c_i;c_o,s)\} & , n > 1 \end{cases}$$

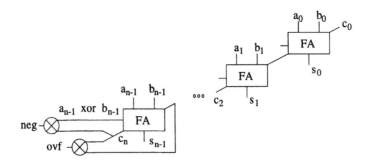

Figure 3.12: Circuit of the ripple carry adder $RCA_n$

### 3.7.3   The Carry Look-ahead Adder (CLA)

The carry look-ahead adder $CLA_n$ speeds up the computation of the carry bits $c_i$ by resolving their recursion formula 3.1. The implementation of $CLA_n$ presented in [Weg87, KP95] achieves cost linear in $n$ and delay logarithmic in $n$. The core of this circuit, as we will see shortly, is a fast circuit for parallel prefix computation.

**Definition 3.10 (Parallel Prefix)** *Let* $\circ : M \times M \rightarrow M$ *be a dyadic function. Its $n$-fold parallel prefix function* $PP_\circ(n) : M^n \rightarrow M^n$ *maps $n$ inputs $x_1, \cdots, x_n$ into $n$ results $y_1, \cdots, y_n$ with $y_i = x_1 \circ \cdots \circ x_i$.*

The circuit recursively defined in figure 3.13 is a fast and cost-effective implementation of the parallel prefix computation of any associative dyadic function $\circ$, at least if $n$ is a power of two. Let $G_\circ$ be the circuit implementing the function $\circ$. The following formulae then estimate the cost and the delay of the parallel prefix circuit $PP_\circ(n)$ in multiples of the cost and the delay of $G_\circ$.

$$
\begin{aligned}
C_{PP}(1) &= 0 \\
C_{PP}(n) &= C_{PP}(n/2) + n - 1 \\
D_{PP}(1) &= 0 \\
D_{PP}(n) &\leq D_{PP}(n/2) + 2
\end{aligned}
$$

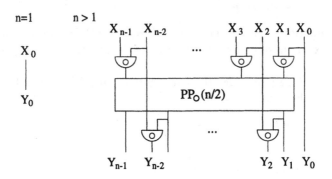

Figure 3.13: The recursive specification of an n-fold parallel prefix circuit of the function $\circ$; n is a power of two

Let $a, b \in \{0, 1\}^n$ and $cin \in \{0, 1\}$ be the inputs of the adder. For $0 \leq i < n$ we define functions $g_{i,i}, p_{i,i} : \{0, 1\}^{2n+1} \rightarrow \{0, 1\}$ with

$$
\begin{aligned}
g_{0,0}(a, b, cin) &= \begin{cases} 1 & \text{if } a_0 + b_0 + cin \geq 2 \\ 0 & \text{otherwise} \end{cases} \\
&= (a_0 \otimes b_0) \wedge cin \vee (a_0 \wedge b_0) \\
g_{i,i}(a, b, cin) &= \left. \begin{cases} 1 & \text{if } a_i + b_i \geq 2 \\ 0 & \text{otherwise} \end{cases} \right\} = a_i \wedge b_i
\end{aligned}
$$

$$p_{i,i}(a, b, cin) = \left\{ \begin{array}{ll} 1 & \text{if } a_i + b_i = 1 \\ 0 & \text{otherwise} \end{array} \right\} = a_i \otimes b_i$$

For the addition of $a, b$ and $cin$, $g_{i,i}(a, b, cin)$ indicates that the position $i$ generates a carry, and $p_{i,i}(a, b, cin)$ indicates that the position $i$ propagates a carry. The following recursion formulae define for any $0 \leq i \leq k < j < n$ functions $g_{i,j}, p_{i,j} : \{0, 1\}^{2n+1} \to \{0, 1\}$. They indicate, that the positions $i$ to $j$ generate respectively propagate a carry.

$$g_{i,j}(a, b, cin) = g_{k+1,j}(a, b, cin) \vee (g_{i,k}(a, b, cin) \wedge p_{k+1,j}(a, b, cin))$$
$$p_{i,j}(a, b, cin) = p_{i,k}(a, b, cin) \wedge p_{k+1,j}(a, b, cin)$$

The sum $s$ and the carries $c_i$ of the numbers $a, b$ and $cin$ can then be expressed as:

$$c_i = g_{0,i-1}$$
$$s_i = p_{i,i} \otimes g_{0,i-1}$$

Consequently, the $CLA_n$ adder first generates all values $g_{i,i}(a, b, cin)$ and $p_{i,i}(a, b, cin)$. It then computes all $g_{0,i}(a, b, cin)$, $p_{0,i}(a, b, cin)$ and in the third step, the adder generates the result $s$ and the flags $ovf$ and $neg$.

For the second step, we have to compute the parallel prefix function $PP_o(n)$ : $\{0, 1\}^{2n} \to \{0, 1\}^{2n}$ for function $o : \{0, 1\}^4 \to \{0, 1\}^2$, with

$$o(x_1, x_0, y_1, y_0) = (x_1 \vee (y_1 \wedge x_0), x_0 \wedge y_0).$$

Since this function $o$ is associative (see [Weg87, KP95]), we can use the fast parallel prefix circuit of figure 3.13 to compute the generate and propagate signals $(g_{0,i}, p_{0,i})$.

This adder design provides the carry bits $c_n$ and $c_{n-1}$ directly. Thus, the mechanism of theorem 3.2 can be used to detect the overflow:

$$ovf = c_n \otimes c_{n-1} = g_{0,n-1} \otimes g_{0,n-2}$$
$$neg = c_n \otimes a_{n-1} \otimes b_{n-1} = c_n \otimes p_{n-1,n-1}$$

Cost and delay of the carry look-ahead adder (figure 3.14) then run at

$$C_{CLA}(n) = C_{PP}(n) \cdot (C_{or} + 2 \cdot C_{and}) + (2n + 2) \cdot C_{xor}$$
$$+ (n + 1) \cdot C_{and} + C_{or}$$
$$D_{CLA}(n) = D_{PP}(n) \cdot (D_{and} + D_{or}) + 2 \cdot D_{xor} + D_{and} + D_{or}$$

### 3.7.4 The Conditional Sum Adder (CSA)

In order to derive the conditional sum implementation of an n-bit adder, we first introduce a circuit $A2(n)$ which for $a, b \in \{0, 1\}^n$ generates the two sums $s^0, s^1 \in \{0, 1\}^{n+1}$ with

$$<s^0> = <a> + <b> + 0$$
$$<s^1> = <a> + <b> + 1.$$

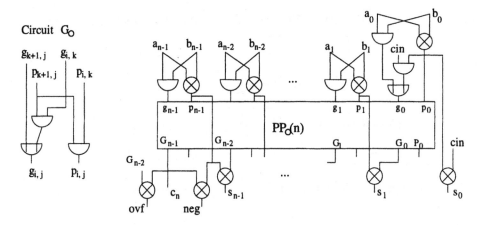

Figure 3.14: n-bit carry look-ahead adder $CLA_n$. We use the following abbreviations: $g_i = g_{i,i}$, $p_i = p_{i,i}$, $G_i = g_{0,i}$ and $P_i = p_{0,i}$.

| $a_0$ | $b_0$ | $s_1^0$ | $s_0^0$ | $s_1^1$ | $s_0^1$ |
|------|------|------|------|------|------|
| 0 | 0 | 0 | 0 | 0 | 1 |
| 0 | 1 | 0 | 1 | 1 | 0 |
| 1 | 0 | 0 | 1 | 1 | 0 |
| 1 | 1 | 1 | 0 | 1 | 1 |

$$s_1^0 = a_0 \wedge b_0, \quad s_0^0 = a_0 \otimes b_0$$
$$s_1^1 = a_0 \vee b_0, \quad s_0^1 = \overline{a_0 \otimes b_0}$$

Table 3.1: Functionality and formulae of circuit $A2(1)$

Table 3.1 describes the functionality of circuit $A2(1)$ and lists formulae for the outputs of $A2(1)$. For $n > 1$, the design of circuit $A2(n)$ is based on the "divide and conquer" principle. Let $m = \lceil n/2 \rceil$ and $k = \lfloor n/2 \rfloor$. The input vectors $a, b \in \{0, 1\}^n$ are then split into vectors $(a_{n-1}, \ldots, a_m)$, $(a_{m-1}, \ldots, a_n)$, $(b_{m-1}, \ldots, b_m)$ and $(b_{m-1}, \ldots, b_n)$ and circuit $A2(n)$ is composed as indicated in figure 3.15. Thus, cost and delay of the n-bit $A2$ adder are:

$$C_{A2}(n) = \begin{cases} C_{xor} + C_{xnor} + C_{and} + C_{or} & n = 1 \\ C_{A2}(k) + C_{A2}(m) + 2 \cdot C_{mux}(k + 1) & n > 1 \end{cases}$$

$$D_{A2}(n) = \begin{cases} \max\{D_{xor}, D_{xnor}, D_{and}, D_{or}\} & n = 1 \\ D_{A2}(m) + D_{mux}(k + 1) & n > 1 \end{cases}$$

The $A2(n)$ adder is the core of the conditional sum implementation of an n-bit adder. In the final step, the conditional sum adder then selects the proper sum – either $s^0$ or $s^1$ – based on the input carry $c_0$. The flag $neg$ is generated as usual

$$neg = c_n \otimes a_{n-1} \otimes b_{n-1}.$$

However, in the conditional sum adder $CSA_n$, most of the carry bits are not directly accessible. Thus, $(n + 1)$-bit arithmetic must be used in order to detect

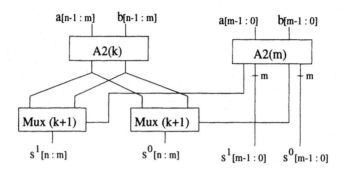

Figure 3.15: Circuit $A2(n)$, $n > 1$, $m = \lceil n/2 \rceil$, $k = \lfloor n/2 \rfloor$

the overflow. According to theorem 3.3, the flag $ovf$ can be generated as

$$ovf = a_{n-1} \otimes b_{n-1} \otimes c_n \otimes s_{n-1} = neg \otimes s_{n-1}.$$

Once again, the expression $a_{n-1} \otimes b_{n-1}$ need not be generated extra, but can be taken from circuit $A2(n)$. Thus the conditional sum adder $CSA_n$ (figure 3.16) has the following cost and delay:

$$C_{CSA}(n) = C_{A2}(n) + C_{mux}(n+1) + 2 \cdot C_{xor}$$
$$D_{CSA}(n) = D_{A2}(n) + D_{mux}(n+1) + 2 \cdot D_{xor}.$$

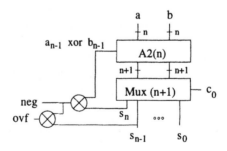

Figure 3.16: Circuit of the conditional sum adder $CSA_n$

## 3.7.5   The Subtraction

**Definition 3.11** *An n-bit subtraction circuit computes the function* $-_n : \{0, 1\}^{2n} \rightarrow \{0, 1\}^{n+2}$,

$$-_n(a_{n-1}, \ldots, a_0, b_{n-1}, \ldots, b_0) = (l, o, s_{n-1}, \ldots, s_0).$$

*If $[a] - [b] \in T_n$, then $o = 0$ and $[s_{n-1}, \ldots, s_0] = [a] - [b]$; otherwise $o = 1$. The flag $o$ indicates an overflow. The flag $l = 1$ if $[a] < [b]$ and $l = 0$ otherwise.*

According to theorem 3.1, $[a] - [b] = [a] + [\bar{b}] + 1$. To implement the n-bit subtraction circuit, we just extend the circuit of a two's complement adder by $n$ inverters and tie *cin* to 1. The signals $o$ and $l$ then correspond to the overflow signal *ovf* and the signal *neg*.

### 3.7.6   The Arithmetic Unit

An arithmetic unit implements both, addition and subtraction; an additional control signal *sub* specifies the operation to be performed.

**Definition 3.12 (Arithmetic Unit)** *An n-bit arithmetic unit computes the function $\pm_n : \{0, 1\}^{2n+1} \to \{0, 1\}^{n+2}$,*

$$\pm_n(a_{n-1}, \ldots, a_0, b_{n-1}, \ldots, b_0, sub) = (neg, ovf, s_{n-1}, \ldots, s_0)$$

*with*

$$[s_n, s_{n-1}, \ldots, s_0] = \begin{cases} [a] + [b] & \text{if } sub = 0 \\ [a] - [b] = [a] + [\bar{b}] + 1 & \text{if } sub = 1, \end{cases}$$

*with $ovf = 0$ if $[s_n, \ldots, s_0] \in T_n$ and $ovf = 1$ otherwise, and with $neg = 1$ if $[s_n, \ldots, s_0] < 0$ and $neg = 0$ otherwise.*

Let be $a, b, d \in \{0, 1\}^n$ with $d_i = b_i \otimes sub$, then $d = b$ for $sub = 0$ and $d = \bar{b}$ for $sub = 1$. We therefore can express the output of the arithmetic unit as

$$(neg, ovf, s_n, \ldots, s_0) = +_n(a, d, sub).$$

Thus, extending the circuit of an $n$-bit adder by $n$ XOR gates as indicated in figure 3.17 yields the circuit of an $n$-bit arithmetic unit. Depending on the implementation of the adder, cost and delay of the arithmetic unit $AU_n$ run at:

$$C_{AU}(n) = C_{xor}(n) + \begin{cases} C_{CLA}(n) & \text{, carry look-ahead} \\ C_{RCA}(n) & \text{, ripple carry} \\ C_{CSA}(n) & \text{, conditional sum} \end{cases}$$

$$D_{AU}(n) = D_{xor}(n) + \begin{cases} D_{CLA}(n) & \text{, carry look-ahead} \\ D_{RCA}(n) & \text{, ripple carry} \\ D_{CSA}(n) & \text{, conditional sum} \end{cases}$$

## 3.8   Cyclic Shifters

**Definition 3.13 (Cyclic Shifter)** *Let $n = 2^m$. An n-bit cyclic left shifter is a circuit which computes the function $cl : \{0, 1\}^{n+m} \to \{0, 1\}^n$,*

$cl(a_{n-1}, \ldots, a_0, b_{m-1}, \ldots, b_0) = (a_{n-\langle b \rangle - 1}, \ldots, a_0, a_{n-1}, \ldots, a_{n-\langle b \rangle})$.

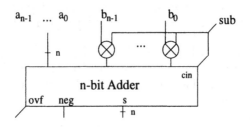

Figure 3.17: Circuit of an n-bit arithmetic unit $AU_n$

An n-bit cyclic right shifter is a circuit which computes the function $cr$ : $\{0, 1\}^{n+m} \rightarrow \{0, 1\}^n$, $cr(a_{n-1}, \ldots, a_0, b_{m-1}, \ldots, b_0) = (c_{n-1}, \ldots, c_0)$, with $c = (a_{<b>-1}, \ldots, a_0, a_{n-1}, \ldots, a_{<b>})$.

We first construct an n-bit cyclic left shifter and then show how that circuit can be used to perform cyclic right shifts.

## 3.8.1   A Cyclic Left Shifter

Let $CLS_{n,i}$ be the circuit which computes the function $\{0, 1\}^{n+1} \rightarrow \{0, 1\}^n$, $(a_{n-1}, \ldots, a_0, s) \mapsto (c_{n-1}, \ldots, c_0)$ with $c = a$ if $s = 0$ and $c = cl(a, bin_m(i))$ otherwise.

This means, that circuit $CLS_{n,i}$ cyclicly shifts input $a$ $i$ bits to the left iff $s = 1$. It can be implemented by an n-bit multiplexer with the data inputs $a$ and $(a_{n-i-1}, \ldots, a_0, a_{n-1}, \ldots, a_{n-i})$, and with the select signal $s$ (figure 3.18).

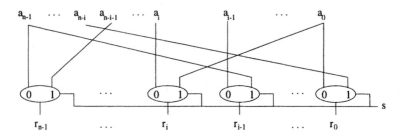

Figure 3.18: Circuit $CLS_{n,i}$

Let $a \in \{0, 1\}^n$, $b \in \{0, 1\}^m$ and $n = 2^m$. Cascading the circuits $CLS_{n,2^0}$, $CLS_{n,2^1}, \ldots, CLS_{n,2^{m-1}}$ as indicated in figure 3.19 yields an n-bit cyclic left shifter, a so called *Barrel shifter*. Its cost and delay run at

$$C_{CLS(n)} = \log n \cdot C_{mux}(n)$$
$$D_{CLS}(n) = \log n \cdot D_{mux}(n).$$

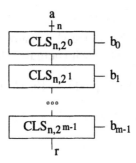

Figure 3.19: Circuit of an n-bit cyclic left shifter $CLS(n)$

## 3.8.2  Cyclic Right Shifts

**Theorem 3.4** *Let* $n = 2^m$, $a \in \{0, 1\}^n$, $b, d \in \{0, 1\}^m$ *and* $x \in \{0, 1\}$ *such that* $<x, d> = <\bar{b}> + 1$. *A cyclic right shift over* $<b>$ *bits then equals a cyclic left shift over* $<d>$ *bits, i.e.,* $cr(a, b) = cl(a, d)$.

**Proof:**

1.  Let $<b> = 0$, then $<\bar{b}> + 1 = <1, 0^m>$ and $<d> = 0$. A cyclic left shift and a cyclic right shift of input $a$ over 0 bits both yield output $a$.

2.  Let $p := <b> > 0$ and $q := <d>$. The results of the left shift

$$cl(a, d) = (a_{n-q-1}, \ldots, a_0, a_{n-1}, \ldots, a_{n-q})$$

    and of the right shift

$$cr(a, b) = (a_{p-1}, \ldots, a_0, a_{n-1}, \ldots, a_p).$$

    are the same iff $n - q = p$.

    From the definition 3.2 of binary numbers, it can easily be derived that $<b> + <\bar{b}> = 2^m - 1$. For $<b> > 0$, we can then conclude that

$$2^m - 1 \geq <\bar{b}> + 1 = <0, d> = <d> = q.$$

    And therefore, we can express $n - q$ as

$$n - q = \sum_{i=0}^{m-1} 2^i + 1 - \left( \sum_{i=0}^{m-1} \bar{b_i} 2^i + 1 \right) = \sum_{i=0}^{m-1} \underbrace{(1 - \bar{b_i})}_{b_i} 2^i = p \qquad \Box$$

According to theorem 3.4, it is sufficient to invert and increment the shift amount $b$ in order to use a cyclic left shifter to perform cyclic right shifts.

## 3.9   Exercises

3.1 (Fanout restriction and fanout trees) We now extend the basic components by a gate called *buffer buf* which computes the identical function

$$buf : \{0, 1\} \rightarrow \{0, 1\}; \ x \mapsto x$$

at cost $C_{buf} = 1$ and delay $D_{buf} = 1$. The symbol of the buffer $buf$ is given below.

Lets impose a *fanout restriction* of $f \geq 2$, i.e., the output of a basic component can be connected to at most $f$ inputs of basic components. In order to tolerate a larger fanout $m > f$, we use buffers arranged as balanced $f$-trees. This circuit is called *fanout tree*.

1. Define the circuit of a fanout tree $Ftree_{f,m}$ which tolerates a fanout of $\lceil m/f \rceil \cdot f$. The fanout tree has only one input wire.

2. Express the cost and delay of the fanout tree $Ftree_{f,m}$ by recursion formulae.

3.2 Designing decoders with limited fanout.

1. Modify the two decoder circuits $decs_n$ and $decf_n$ of section 3.2 such that the new circuits $Fdecs_{n,f}$ and $Fdecf_{n,f}$ satisfy the fanout restriction for $f = 2$ $(f = 4)$.

2. Compare the cost and the delay of the new decoder circuits. Is decoder $Fdecf_{n,f}$ still cheaper and faster than decoder $Fdecs_{n,f}$?

3.3 Design a fanout bound version $Fenc_{n,f}$ of the encoder circuit $enc_n$ of section 3.4 and derive formulae for its cost and delay.

3.4 (Motorola specific optimization) The Motorola technology provides NAND and NOR gates at the same cost as AND and OR gates but at half the delay (table 2.1 of section 2.4.1).

1. Based on the rules of deMorgan, replace the AND and OR gates of the full adder circuit FA of section 3.7.2 by NAND gates.

2. How does this modification impact the cost and delay of the full adder and of the ripple carry adder $RCA_n$ (section 3.7.2)?

3.5 Under Motorola technology, the circuit $CLA_n$ of the carry look-ahead adder (section 3.7.3) can be sped up as well.

1. Derive the recursion formulae of the cost and delay of the faster carry look-ahead adder $MCLA_n$.

2. Use a C program in order to compare the cost and delay of the two 32-bit adder circuits.

3.6 1. For each of the three adder circuits of section 3.7, i.e., the ripple carry adder $RCA_n$, the carry look-ahead adder $CLA_n$ and the conditional sum adder $CSA_n$, derive circuits $FRCA_{n,f}$, $FCLA_{n,f}$ and $FCSA_{n,f}$ which have a fanout of at most $f = 2$ (respectively $f = 4$).

2. For $n = 32$, compare the cost and delay of the adder circuits under Motorola and Venus technology.

3. Perform the same task for the Motorola specific adder circuits $MRCA_n$ (exercise 3.4) and $MCLA_n$ (exercise 3.5).

# Chapter 4

# Hardwired Control

## 4.1 Representation of the Control

Let $I_1, \ldots, I_\sigma$ be the input signals of the control, i.e., the signals coming into the control from the data paths and let $O_1, \ldots, O_\gamma$ be the output signals of the control, i.e., the signals going out of the control into the data paths. Then, the control is obviously a *Mealy automaton*, i.e., a finite state transducer [HU79, Har64] with an output function $\tau$ which depends on both the current state and the current input symbol. Any input symbol specifies the value of all input signals $I_j$, and any output symbol specifies the value of all output signals $O_j$. A *Moore automaton* is a Mealy automaton whose output function depends only on the current state.

For the high level description of the control of computers one often uses so called finite state diagrams (FSD). Figure 4.1 depicts an example which is almost literally from [HP90]. We are presently interested in the *format* of this figure.

The states of the transducer are depicted as boxes. We will denote the set of these states by $Z$. Each box of a state contains a mnemonic name of the state as well as some instructions from a register transfer language to be executed in that state. These instructions specify the output function of the transducer. The transducer is a Moore automaton if none of these instructions tests an input signal of the control. We will use here the register transfer language $RTL$ specified in appendix A.

Each instruction of a register transfer language specifies in a straightforward way the control signals which should be activated. Therefore, for each output signal $O_j$, the set

$$Z_j = \{z \in Z \mid O_j \text{ is active in state } z\}$$

and its cardinality

$$\nu_j = \#Z_j$$

can directly be determined by inspection of the FSD.

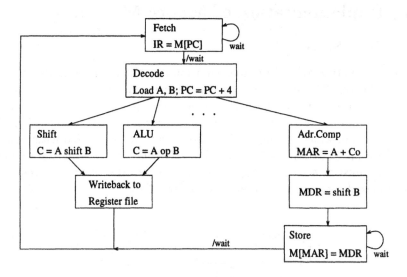

Figure 4.1: Portion of the finite state diagram of the DLX architecture. Tables 6.1 – 6.3 list the DLX instruction set. Figure 6.3 lists the data paths. The whole FSD can be found in figure 6.18.

The next state function of the transducer is determined by the arcs between the boxes and, in case there are several arcs from one box to other boxes, in a somewhat informal way by the mnemonic names of those boxes. We denote by $E$ the set of all pairs of states $(z, z')$ such that the FSD contains an arc from $z$ to $z'$, i.e., the set of edges if we view the FSD as a directed graph.

The mnemonics of states typically refer to the type of instruction that is executed. From the machine language and the FSD one can therefore easily construct for all $(z, z') \in E$ a disjunctive normal form $D(z, z')$ with variables in $I_1, \ldots, I_\sigma$ such that $z'$ is the next state after $z$ iff $D(z, z') = 1$. Many of the formulae $D(z, z')$ are simply the constant 1. We denote by $M(z, z')$ the set of monomials in $D(z, z')$ and by

$$M = \bigcup_{(z,z')\in E} M(z, z') \setminus \{1\}$$

the set of all nontrivial such monomials. We define the *weight* of an arc from $z$ to $z'$ as $\#M(z, z')$. Also, for each state $z'$ we denote by $fanin(z')$ the sum of the weights of all arcs entering state $z'$:

$$fanin(z') = \sum_{(z,z')\in E} \#M(z, z').$$

## 4.2 Implementation of Moore Machines

### 4.2.1 The State

Figure 4.2 depicts a standard implementation of a Moore machine. Let $k = \#Z$ and $\zeta = \lceil \log k \rceil$. The states are coded in binary. The current state is held in a $\zeta$-bit register $\mu$ which is clocked in every cycle. A $\zeta$-bit decoder $dec$ has for each state $z$ an output $S_k$ which is active if register $\mu$ holds (the code of) $z$. The initial state $z_0$ in which execution of an instruction starts is coded with all zeros.

On powerup — external signal $pup$ (PowerUP) is active — the state $\mu$ of the automaton must be set to $z_0$. A $\zeta$-bit AND gate with the output of circuit NSE and $\zeta$ copies of signal $/pup$ as inputs takes care of that.

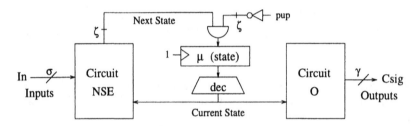

Figure 4.2: Standard implementation of a Moore machine. Circuit $O$ implements the output function of the automaton and circuit $NSE$ computes its next state.

### 4.2.2 Generating the Outputs $O_j$

Outputs are produced by circuit $O$. For each $j$ output $O_j$ is computed by

$$O_j = \bigvee_{z \in Z_j} S_z.$$

The $(\nu_j - 1)$ OR gates for $O_j$ can be arranged as a tree of depth $\lceil \log \nu_j \rceil$. Let $\nu_{max}$ denote the maximal cardinality of the $\gamma$ sets $Z_j$ and let $\nu_{sum}$ denote the sum of all the cardinalities $\nu_j$

$$\nu_{max} = \max\{\nu_j \mid 1 \le j \le \gamma\} \quad ; \quad \nu_{sum} = \sum_{j=1}^{\gamma} \nu_j.$$

Let $OD$ be the circuit consisting of the decoder $dec$ and the circuit $O$. All the inputs of $OD$ come directly from the register $\mu$. Thus,

$$
\begin{aligned}
C_{OD} &= C_{dec}(\zeta) + C_O = C_{dec}(\zeta) + \sum_{i=1}^{\gamma}(\nu_i - 1) \cdot C_{or} \\
&= C_{dec}(\zeta) + (\nu_{sum} - \gamma) \cdot C_{or}
\end{aligned}
$$

$$D_O = \max\{\lceil \log \nu_i \rceil \mid 1 \le i \le \gamma\} \cdot D_{or} = \lceil \log \nu_{max} \rceil \cdot D_{or}$$
$$A_{OD} = D_{OD} = D_{dec}(\zeta) + D_O.$$

### 4.2.3 Computing the Next State

For monomials $m$ we denote by $l(m)$ the number of literals in $m$. By $l_{max}$, we denote the length of the longest monomial $m \in M$, and by $l_{sum}$, we denote the accumulated length of all those monomials $m$

$$l_{max} = \max\{l(m) \mid m \in M\}$$
$$l_{sum} = \sum_{m \in M} l(m).$$

Let circuit $CM$ compute all monomials in $M$ in the following way:

1. Compute $\overline{I_j}$ for each input signal $I_j$.

2. Compute each monomial by a balanced tree of AND gates.

Circuit $CM$ receives the inputs $In$ from the hardware $H$. Let these signals have an accumulated delay of $A(In) = A_H(In)$. Thus,

$$
\begin{aligned}
C_{CM} &= \sigma \cdot C_{inv} + \sum_{m \in M} (l(m) - 1) \cdot C_{and} \\
&= \sigma \cdot C_{inv} + (l_{sum} - \#M) \cdot C_{and} \\
A_{CM} &= A(In) + D_{inv} + \max\{\lceil \log l(m) \rceil \mid m \in M\} \cdot D_{and} \\
&= A(In) + D_{inv} + \lceil \log l_{max} \rceil \cdot D_{and}.
\end{aligned}
$$

For each state $z$ let

$$N(z) = \bigvee_{(z',z) \in E} \bigvee_{m \in M(z',z)} (S_{z'} \wedge m)$$

be the function indicating that $z$ is the next state. A circuit $CN(z)$ computing all these functions from inputs $S_{z'}$ and $M(z',z)$ can be constructed such that

$$
\begin{aligned}
C_{CN(z)} &= \text{fanin}(z) \cdot (C_{and} + C_{or}) - C_{or} \\
D_{CN(z)} &= \lceil \log(\text{fanin}(z)) \rceil \cdot D_{or} + D_{and}.
\end{aligned}
$$

Let $CN$ be a circuit consisting of all the circuits $CN(z)$. Feeding the outputs of circuit $CN$ into a $\zeta$-bit encoder $enc$ yields the (code of the) next state. Note that the encoders constructed in section 3 ignore input 0. In circuit $CN$, we can therefore omit the computation of the $N(z)$ for the state with code $0 \ldots 0$, i.e., for the initial state $z_0$. This also guarantees, that the control automaton switches to state $z_0$ on an undefined transition. Thus, the automaton can not get stuck. It is also a nice coincidence, that $z_0$ tends to be the state with the largest indegree.

Let $\text{fanin}_{max}$ denote the maximal fanin of the $(k-1)$ states $z$ different from the initial state $z_0$, and let $\text{fanin}_{sum}$ denote the accumulated fanin of these $(k-1)$ states

$$\begin{aligned}
\text{fanin}_{max} &= \max\{\text{fanin}(z) \mid z \neq z_0\} \\
\text{fanin}_{sum} &= \sum_{z \neq z_0} \text{fanin}(z).
\end{aligned}$$

Cost and delay of circuit $CN$ then run at

$$\begin{aligned}
C_{CN} &= \sum_{z \neq z_0} C_{CN(z)} = \text{fanin}_{sum}(C_{and} + C_{or}) - (k-1)C_{or} \\
D_{CN} &= \max\{D_{CN(z)} \mid z \neq z_0\} = \lceil \log(\text{fanin}_{max}) \rceil \cdot D_{or} + D_{and}.
\end{aligned}$$

Let $NSE$ be the circuit consisting of circuits $CM$, $CN$ and encoder $enc$. Since $NSE$ receives inputs $In$ from the hardware and the encoded state from the decoder of circuit $OD$, its cost and accumulated delay can be expressed as

$$\begin{aligned}
C_{NSE} &= C_{CM} + C_{CN} + C_{enc}(\zeta) \\
A_{NSE} &= \max\{D_{dec}(\zeta), A_{CM}\} + D_{CN} + D_{enc}(\zeta).
\end{aligned}$$

Thus, cost and delay of all circuits in the above construction depend in a simple way on the parameters listed in table 4.1. These parameters can easily be determined from the FSD and the coding of the instruction set. The C-routines corresponding to the cost and the delay formulae of this and the following hardwired implementations of the control automaton are captured in the C-module listed in appendix B.4.

| Symbol | Meaning |
|--------|---------|
| $\sigma$ | # inputs $I_j$ of the automaton |
| $\gamma$ | # output signals $O_j$ of the automaton |
| $k$ | # states of the FSD (automaton) |
| $\zeta$ | $\zeta = \lceil \log k \rceil$ |
| $\nu_{max}$ | maximal frequency of a control signal in FSD |
| $\nu_{sum}$ | accumulated frequency of all control signals |
| $\#M$ | # monomials $m \in M$ (nontrivial) |
| $l_{max}$ | length of longest monomial $m \in M$ |
| $l_{sum}$ | accumulated length of all monomials $m \in M$ |
| $\text{fanin}_{max}$ | maximal fanin of nodes ($\neq$ fetch) in FSD |
| $\text{fanin}_{sum}$ | accumulated fanin of nodes ($\neq$ fetch) in FSD |

Table 4.1: Parameters relevant for cost and delay of the control automaton

### 4.2.4   Cost and Delay

The Moore machine consists of the circuits $NSE$ and $OD$, the state register, $\zeta$ AND gates and one inverter. Figure 4.3 depicts the structure of this implementation. Its cost amounts to

$$C_{Moore} = C_{NSE} + C_{OD} + C_{ff}(\zeta) + \zeta \cdot C_{and} + C_{inv}.$$

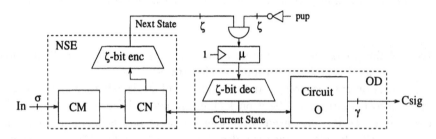

Figure 4.3: Implementation of a Moore machine

The accumulated delay $A_{Moore}(Csig)$ measures the time spent at the beginning of each cycle in order to determine the current control signals. For our realization of the Moore machine, that equals the delay of circuit $OD$. $T_{Moore}$ measures the time required for the update of the registers in the control, and it therefore specifies the minimal cycle time of the control logic. The external signal $pup$ is already valid at the start of a new cycle. Since the delay of circuit NSE is much larger than $D_{inv}$, $T_{Moore}$ equals the accumulated delay of circuit $NSE$ plus the delay of the AND gates and the time $\Delta = D_{ff} + \delta$ for clocking the registers.

$$A_{Moore}(Csig) = A_{OD}$$
$$T_{Moore} = A_{NSE} + D_{and} + \Delta.$$

### 4.2.5   Unary Coding

Coding the current state in unary enlarges the state register $\mu$ and the powerup circuit, but the $\zeta$-bit decoder and encoder can be dropped. However, we now have to ensure that transitions to state $z_0$ are performed in a correct manner. Circuit $CN$ computes signals $N(z)$ only for $z \neq z_0$, but signal $N(z_0)$ must be set iff $N(z) = 0$ for all $z \neq z_0$. A (k-1)-bit zero tester can perform this task.

On powerup ($pup = 1$), the state register $\mu$ must be set to $z_0$, i.e., $\mu = 0^{k-1}1$. This can be realized by $k-1$ AND gates, an OR gate and an inverter, as indicated in figure 4.4. Thus, the cost of the Moore machine with unary coded states runs at

$$C_{uMoore} = C_{Moore} + (k - \zeta) \cdot C_{ff} + C_{zero}(k - 1) + C_{or}$$
$$+ (k - \zeta - 1) \cdot C_{and} - C_{enc}(\zeta) - C_{dec}(\zeta).$$

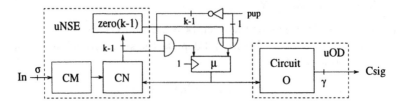

Figure 4.4: Implementation of a Moore machine with states coded in unary

We add the prefix "u" to a circuit's name, if it is necessary to indicate that the circuit is used together with unary coded states.

It depends on the technology and the number of states whether unary coding results in a cost improvement (section 4.4), but it definitely speeds up the control. The circuit $uOD$ does not need to decode the state and is therefore faster by the delay of the decoder. Since the decoder and the circuit $CM$ are traversed in parallel, the decoder may not lie on the critical path of $uNSE$. With unary coding, the modified circuits $uOD$ and $uNSE$ have the following accumulated delays:

$$
\begin{aligned}
A_{uOD} &= A_{OD} - D_{dec}(\varsigma) = D_O \\
A_{uNSE} &= A(In) + D_{CM} + D_{CN} + D_{zero}(k-1).
\end{aligned}
$$

Thus, in the Moore automaton with unary coded state, the computation of the outputs $Csig$ requires

$$
A_{uMoore}(Csig) = A_{uOD}
$$

gate delays, and the cycle time of the automaton runs at

$$
T_{uMoore} = A_{uNSE} + \max\{D_{and}, D_{or}\} + \Delta.
$$

## 4.3 Mealy Implementation

In the Moore implementation of the previous section, each cycle starts with the computation of the control signals. Thus, the output logic $OD$ lies almost always on the critical path of the hardware. However, the control signals depend on the current state but not on the current input signals. In our second implementation, we therefore precompute the control signals of the next state and store them in a register $\mu_O$. At the beginning of a cycle, the control signals are then valid immediately. That circumvents the major drawback of our first approach.

Even after this modification, the control is still a Moore automaton. Its state is now coded in the registers $\mu$ and $\mu_O$. Without register $\mu_O$, our second implementation is obviously of the form of a Mealy automaton, i.e., the outputs depend on the current state *and* on the current input. In the following, we therefore refer to the second implementation as the *Mealy implementation of a Moore automaton*, or for short, as the *Mealy implementation*.

### 4.3.1  Transformations

A simple modification transforms the Moore implementation of figure 4.3 in the Mealy implementation shown in figure 4.5. The output logic reads the next state instead of the current state and saves the output signals in register $\mu_O$. That requires an additional $\gamma$-bit register. The $\zeta$-bit decoder of the circuit $OD$ is still necessary, but it is now exclusively used by circuit $NSD$ which computes the next state of the automaton.

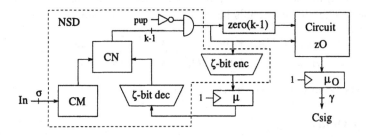

Figure 4.5: Mealy implementation with binary coded state

**Powerup Mechanism**

On powerup ($pup = 1$), both registers must be initialized. Since the initial state $z_0$ is coded with all bits zero, register $\mu$ is to be cleared. Register $\mu_O$ gets the output vector corresponding to state $z_0$. A simple powerup mechanism just forces the outputs of circuit $CN$ to zero on powerup. The encoder and the output logic then get the proper state. This can be done with $k - 1$ AND gates and one inverter (figure 4.5). Thus, the computation of the next state basically remains the same and still requires time

$$A_{NSD} = A_{CN} + D_{and} + D_{enc}(\zeta).$$

**Generating Outputs $O_j$**

For all $z \in Z$, let signal $N'(z)$ indicate that state $z$ is the next state of the automaton. The powerup circuit provides the signals $N'(z_i)$ for all $i \geq 1$ to the output circuit, and a $(k - 1)$-bit zero tester generates signal $N'(z_0)$ from those inputs.

Thus, circuit $O$ of the Moore implementation could now be used in order to generate the output signals $O_j$. Their accumulated delay would then run at $A_{CN} + D_{and} + D_{zero}(k - 1) + D_O$, but the following circuit $zO$ is much faster.

An output signal $O_j$ falls into one of two classes, either it is active during state $z_0$ (i.e., $z_0 \in Z_j$) or it is not (i.e., $z_0 \notin Z_j$). Signals of the latter type are still generated as

$$O_j = \bigvee_{z \in Z_j} N'(z) = \bigvee_{z \in Z_j \setminus \{z_0\}} N'(z).$$

For a signal $O_j$ with $z_0 \in Z_j$, we modify the computation as follows

$$O_j = \left( \bigvee_{z \in Z_j \setminus \{z_0\}} N'(z) \right) \vee N'(z_0).$$

Thus, the computation of the output signals and the test for zero can largely be overlapped. The output signals have an accumulated delay of at most

$$
\begin{aligned}
A_{zO} &= A_{CN} + D_{and} + \max\{D_{zero}(k-1), \lceil \log \nu_{max} \rceil \cdot D_{or}\} + D_{or} \\
&= A_{CN} + D_{and} + \max\{D_{zero}(k-1), D_O\} + D_{or}.
\end{aligned}
$$

The circuits $O$ and $zO$ have the same cost.

### Cost and Delay

After this transformation, the control automaton requires an additional $\gamma$-bit register, an $(k-1)$-bit zero tester and additional $(k-1-\zeta)$ AND gates. Thus, the cost of the Mealy implementation runs at

$$C_{Mealy} = C_{Moore} + C_{ff}(\gamma) + C_{zero}(k-1) + (k-1-\zeta) \cdot C_{and}.$$

The computation of the next state remains the same. Only the time required to generate the control signals varies. These signals are now precomputed in the previous cycle. Thus, the signal $Csig$ are valid at the beginning of a cycle. This modification also impacts the minimal cycle time $T_{Mealy}$ of the automaton, because two register must be updated.

$$
\begin{aligned}
A_{NSD} &= A_{CN} + D_{and} + D_{enc}(\zeta) \\
A_{zO} &= A_{CN} + D_{and} + \max\{D_{zero}(k-1), D_O\} + D_{or} \\
T_{Mealy} &= \max\{A_{NSD}, A_{zO}\} + \Delta \\
A_{Mealy}(Csig) &= 0
\end{aligned}
$$

This Mealy implementation has the same functionality and roughly the same cycle time as the Moore implementation introduced in the previous section.

**Theorem 4.1** *If $D_{nor} \leq D_{or}$, if binary coding the states of the automaton, and if using a standard encoder (i.e., an encoder which computes the function of definition 3.5 (section 3) and which is build from OR gates with two inputs and one output) to encode the states, then the cycle time of the Mealy implementation described above is at most $2 \cdot D_{or}$ gate delays longer than the cycle time of the Moore implementation of section 4.2.*

**Proof:**

In our Moore and Mealy implementations, the update of the control registers requires time:

$$\begin{aligned}
T_{Moore} &= A_{NSE} + D_{and} + \Delta \\
&= \max\{D_{dec}(\zeta),\, A(In) + D_{CM}\} + D_{CN} \\
&\quad + D_{enc}(\zeta) + D_{and} + \Delta \\
T_{Mealy} &= \max\{A_{NSD},\, A_{zO}\} + \Delta \\
&= \max\{D_{dec}(\zeta),\, A(In) + D_{CM}\} + D_{CN} + D_{and} \\
&\quad + \max\{D_{enc}(\zeta),\, \max\{D_{zero}(k-1),\, D_O\} + D_{or}\} + \Delta
\end{aligned}$$

The control automaton receives input signals $In$. Their accumulated delay $A(In)$ usually depends on the delay of the control signals. In the Mealy implementation, the control signals are valid immediately at the start of the cycle, but that is not the case for the Moore implementation. Thus, the difference of the two cycle times is at most

$$\begin{aligned}
T_{Mealy} - T_{Moore} &\geq \max\{0,\, \max\{D_{zero}(k-1),\, D_O\} \\
&\qquad\qquad \overset{!}{+} D_{or} - D_{enc}(\zeta)\} \leq 2D_{or}.
\end{aligned}$$

Hence, it is sufficient to show that the delay of the $(k-1)$-bit zero tester and of the circuit $O$ are at most $D_{or}$ delays longer than the delay of a standard $\zeta$-bit encoder.

Since the NOR gate is not slower than the OR gate, the delay of the $(k-1)$-bit zero tester (section 3.5) is at most

$$\begin{aligned}
D_{zero}(k-1) &= (\lceil \log(k-1) \rceil - 1) \cdot D_{or} + D_{nor} \\
&\leq (\lceil \log k \rceil - 1) \cdot D_{or} + D_{or} = \zeta \cdot D_{or}.
\end{aligned}$$

The circuit $O$ has delay

$$D_O = \max\{\lceil \log \nu_i \rceil \mid 1 \leq i \leq \gamma\} \cdot D_{or} = \lceil \log \nu_{max} \rceil \cdot D_{or};$$

$\nu_i$ is the cardinality of $Z_i \subseteq Z$, the set of states in which control signal $O_i$ is active. At worst, $\nu_{max} = k$, and

$$D_O \leq \lceil \log k \rceil \cdot D_{or} = \zeta \cdot D_{or}.$$

Hence, $\zeta \cdot D_{or}$ is an upper bound for the delay of circuit $O$ and for the delay of the zero tester.

It can be shown, that the standard $\zeta$-bit encoder of definition 3.5 (section 3) has a delay of at most $(\zeta - 1)D_{or}$. Thus,

$$\max\{D_{zero}(k-1),\, D_O\} \leq \zeta \cdot D_{or} \leq D_{enc}(\zeta) + D_{or},$$

and the Mealy implementation of the automaton is at most $2 \cdot D_{or}$ delays slower than the Moore implementation.                    $\square$

## 4.3.2 Unary Coding

Even with the modification mentioned above, the control might still lie on the critical path of the hardware, but unary coding of the current state could speed up the Mealy implementation. This modification enlarges the state register, but there is no need for an encoder or decoder anymore. The cost of the Mealy implementation with unary coded states (figure 4.6) therefore runs at

$$C_{uMealy} = C_{Mealy} + (k - \varsigma) \cdot C_{ff} - C_{enc}(\varsigma) - C_{dec}(\varsigma).$$

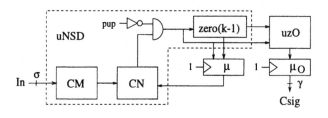

Figure 4.6: Mealy implementation with states coded in unary

Once again, we add the prefix "u" to a circuit's name in order to indicate that this circuit is used together with unary coded states. Unary coding may also impact the cycle time of the control unit. In the Mealy implementation, the encoder and the output circuit with zero tester are traversed in parallel. The same holds for the decoder and the circuit $CM$. When coding the states in unary, the Mealy implementation therefore has the following delays:

$$
\begin{aligned}
A_{uNSD} &= A_{CM} + D_{CN} + D_{and} + D_{zero}(k - 1) \\
A_{uzO} &= A_{CM} + D_{CN} + D_{and} + \max\{D_{zero}(k - 1), D_O\} + D_{or} \\
T_{uMealy} &= \max\{A_{uNSD}, A_{uzO}\} + \Delta \\
&= A_{CM} + D_{CN} + D_{and} + \max\{D_{zero}(k - 1), D_O\} \\
&\quad + D_{or} + \Delta \\
A_{uMealy}(Csig) &= 0
\end{aligned}
$$

Since the Mealy implementation with binary coded states has the cycle time

$$
\begin{aligned}
T_{Mealy} &= \max\{A_{NSD}, A_{zO}\} + \Delta \\
&= \max\{D_{dec}(\varsigma), A_{CM}\} + D_{CN} + D_{and} + \max\{D_{enc}(\varsigma), \\
&\quad \max\{D_{zero}(k - 1), D_O\} + D_{or}\} + \Delta,
\end{aligned}
$$

and since the zero tester is slower than the encoder, unary coding only improves the run time of the Mealy automaton if $A(In) + D_{CM} < D_{dec}(\varsigma)$, i.e., if all monomials in $M$ are short and if the inputs $In$ of the automaton have a short delay.

## 4.4   Cost Impact of Binary Coding

The states of the FSD can be coded in binary or in unary. On Mealy and Moore implementation of the control automata, that results in a different cost difference

$$
\begin{aligned}
\Delta_{Moore}(k,\zeta) \ := \ & C_{uMoore} - C_{Moore} \\
= \ & (k-\zeta) \cdot C_{ff} - C_{dec}(\zeta) - C_{enc}(\zeta) \\
& + C_{zero}(k-1) + (k-\zeta-1) \cdot C_{and} + C_{or} \\
\Delta_{Mealy}(k,\zeta) \ := \ & C_{uMealy} - C_{Mealy} \\
= \ & (k-\zeta) \cdot C_{ff} - C_{dec}(\zeta) - C_{enc}(\zeta).
\end{aligned}
$$

Unary coding saves the cost of a $\zeta$-bit decoder and encoder but requires at least $k-\zeta$ additional flipflops. It depends on the number of states $k$ and the technology whether unary or binary coding is cheaper. We first analyze the more complex equation of the Moore implementation and then transfer the results to the Mealy implementation.

### 4.4.1   Impact on the Moore Implementation

The encoder, the decoder and the zero tester of chapter 3 are built from inverters, AND gates, OR gates, and one NOR gate. Independent of the technology, the costs of these gates usually obey the relation

$$
\begin{aligned}
C_{and} = C_{or} = 2 \cdot C_{inv} \\
C_{nor} \leq C_{or}.
\end{aligned}
\tag{4.1}
$$

Hence, the cost of the decoder, encoder and zero tester can be expressed as a product of the parameter $C_{and}$ and a technology independent function

$$
C_{dec}(\zeta) + C_{enc}(\zeta) \ = \ g(\zeta) \cdot C_{and}
$$

$$
C_{zero}(k-1) \ = \ (k-3) \cdot C_{or} + C_{nor} \ \leq \ (k-2) \cdot C_{and}.
$$

Thus, binary coding is cheaper than unary coding for the Moore implementation if

$$
\Delta_{Moore}(k,\zeta) \ = \ (k-\zeta) \cdot C_{ff} - g(\zeta) \cdot C_{and} + (2k-\zeta-2) \cdot C_{and} \ > \ 0,
$$

i.e., if

$$
f(k,\zeta) := \frac{g(\zeta) - 2k + \zeta + 2}{k-\zeta} \ < \ \frac{C_{ff}}{C_{and}}.
$$

This equation compares the impact of technology and the impact of the size of the FSD. Since $\zeta = \lceil \log k \rceil$, the value of $k$ must lie within the range $[2^{\zeta-1}+1,\, 2^{\zeta}]$. Let

$$
\begin{aligned}
f_{min}(\zeta) \ &:= \ f(2^{\zeta}, \zeta) \\
f_{max}(\zeta) \ &:= \ f(2^{\zeta-1}+1, \zeta),
\end{aligned}
$$

then for all $k$ and $\zeta$

$$f_{min}(\zeta) \leq f(k,\zeta) \leq f_{max}(\zeta).$$

Figure 4.7 depicts the graphs of the two functions $f_{min}$ and $f_{max}$; both have a peak at $\zeta = 3$ and level out for large values $\zeta > 10$.

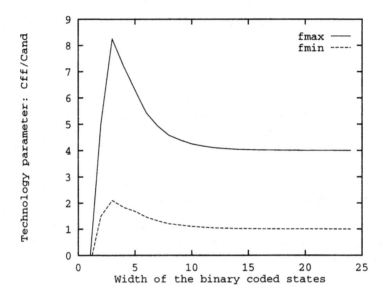

Figure 4.7: Graph of $f_{min}(\zeta)$ and $f_{max}(\zeta)$ for the Moore implementation

For a fixed value $\zeta$, binary coding is cheaper for all $k$ if the relative cost $C_{ff}/C_{and}$ exceeds the value $f_{max}(\zeta)$. On the other hand, unary coding is cheaper for all $k$ if the relative cost $C_{ff}/C_{and}$ is less than $f_{min}(\zeta)$. In any other case, there exists a break-even point

$$k_{min}(\zeta) = \left\lceil \frac{(g(\zeta)+2) \cdot C_{and} + \zeta \cdot (C_{ff} + C_{and})}{C_{ff} + 2 \cdot C_{and}} \right\rceil.$$

for every $\zeta$. Past this point, binary coding becomes profitable.

Under Venus technology with $C_{ff}/C_{and} = 6$, the relative cost of a flipflop exceed $f_{max}(\zeta)$ for $\zeta \geq 6$; and the exact break-even points of binary coding (table 4.2) even indicate that for more than 18 states, binary coding is cheaper than unary coding.

Motorola technology provides flipflops at the relative cost of $C_{ff}/C_{and} = 4$. Thus, the relative cost of a flipflop lies between the bounds $f_{min}(\zeta)$ and $f_{max}(\zeta)$, and only for large FSDs ($\zeta > 12$), it approaches the value of $f_{max}$. For medium sized FSDs, it truly depends on the size $k$ whether it is cost effective to code the states in binary (see table 4.2).

| $\zeta$ | | 1 | 2 | 3 | 4 | 5 | 6 | 7 | 8 | 9 |
|---------|---|---|---|---|---|---|---|---|---|---|
| Range of $k$ | | 1.. 2 | 3.. 4 | 5.. 8 | 9.. 16 | 17.. 32 | 33.. 64 | 65.. 128 | 129.. 256 | 257.. 512 |
| Moore | MC | 2 | 4 | 7 | 12 | 22 | 40 | 75 | 141 | 274 |
| | V | 2 | 3 | 6 | 10 | 18 | 32 | 58 | 108 | 208 |
| | A | 2 | 4 | 8 | 16 | 30 | 57 | 108 | 208 | 406 |
| Mealy | MC | 2 | 4 | 9 | 16 | 31 | 58 | 109 | 209 | 408 |
| | V | 2 | 4 | 7 | 12 | 23 | 41 | 75 | 142 | 275 |
| | B | 2 | 3 | 6 | 10 | 18 | 32 | 58 | 109 | 209 |

Table 4.2: Break-even points $k_{min}(\zeta)$ of binary coding for the Moore and the Mealy implementation under Motorola (MC) and Venus (V) technology. The technologies A and B provide flipflops at the realive cost of $C_{ff}/C_{and} = 2$ respectively 8.

## 4.4.2 Impact on the Mealy Implementation

For a Mealy implementation, switching from binary to unary coding has a less severe impact on the cost, because the variant with binary coded states already requires the zero tester and the wide powerup circuit. Based on equation 4.1, the cost difference can be expressed as

$$\Delta_{Mealy}(k, \zeta) = (k - \zeta) \cdot C_{ff} - g(\zeta) \cdot C_{and} > 0$$

and thus, for the Mealy implementation, binary coding is cheaper than unary coding if

$$F(k, \zeta) := \frac{g(\zeta)}{k - \zeta} < \frac{C_{ff}}{C_{and}}.$$

In analogy to the previous section, this function is bound by

$$\begin{aligned} F_{min}(\zeta) &:= F(2^\zeta, \zeta) \\ F_{max}(\zeta) &:= F(2^{\zeta-1} + 1, \zeta), \end{aligned}$$

and the break-even points $k_{min}(\zeta)$ of binary coding (table 4.2) can now be expressed as

$$k_{min}(\zeta) = \left\lceil \frac{g(\zeta) \cdot C_{and}}{C_{ff}} \right\rceil + \zeta.$$

The graphs of the two functions $F_{min}$ and $F_{max}$ (figure 4.8) have the same shape but a different scaling than those of the Moore implementation. For large $\zeta$, they now approach a value of the technology parameter $C_{ff}/C_{and}$ which is by 2 higher than before. This indicates a push towards unary coding. Thus, in combination with a Mealy implementation, binary coding of the states is not a realistic option if the relative cost of a flipflop is at most $C_{ff}/C_{and} = 3$.

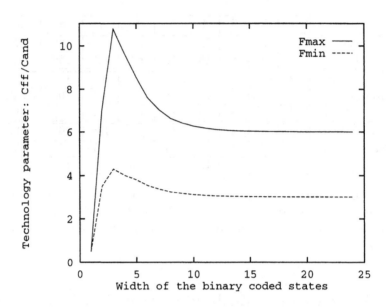

Figure 4.8: Graph of $F_{min}(\zeta)$ and $F_{max}(\zeta)$ for the Mealy implementation

**Theorem 4.2** *For reasonably sized FSDs, i.e., FSDs with 20 to 50 states, the break-even points in table 4.2 indicate the following:*

- *If the relative cost of flipflops is high $C_{ff}/C_{and} \geq 8$, coding the states in binary improves the cost of Mealy and of the Moore implementation.*

- *If the technology provides cheap flipflops with relative cost of at most 2 $(C_{ff}/C_{and} \leq 2)$, binary coding is not a realistic option for either implementation.*

## 4.5  Exercises

4.1 For the Moore and Mealy implementation with states coded in binary, theorem 4.1 states that the cycle time of the Mealy implementation is at most $2 \cdot D_{or}$ gate delays longer than the one of the Moore implementation.

However, the control automaton gets inputs $In$ from the data paths, and the computation of these inputs usually depends on the control signals generated by the automaton. Thus, it is a reasonable assumption, that the accumulated delay of the inputs $In$ obeys the following formula for an $\alpha \geq 0$:

$$A(In) = \left\{ \begin{array}{l} A_{Moore}(Csig) \\ A_{Mealy}(Csig) \end{array} \right. + \alpha$$

Prove under this assumption that

$$T_{Moore} \leq T_{Mealy} + (\lceil \log \nu_{max} \rceil - 1) \cdot D_{or}.$$

4.2 Derive similar theorems for the cycle times $T_{uMoore}$ and $T_{uMealy}$ of the Moore and Mealy implementation with states coded in unary.

# Chapter 5

# Design of a Minimal CPU

By Thomas Grün

In the last chapters, we have introduced the prerequisites for analyzing a computer architecture. Before we start to design the fairly comprehensive DLX machine, we illustrate the design and evaluation process on a minimal CPU derived from the Manchester MARK 1 [SBN82, pp. 107–109].

The MARK 1, built in 1948, was the world's first memory programmable computer. It was designed for testing a new CRT (cathode ray tube) memory technology, called Wiliams tubes [WKT51]. Since the design goal was limited to testing new memory devices, there was no need for a powerful, general purpose instruction set and hence, only a very simple instruction set was used.

## 5.1 Design

The original MARK 1 had only one general purpose 32-bit register, called the accumulator, denoted by ACC. The value stored in the accumulator could either be read from the main memory or written to it. The main memory had a size of 32 words and could be extended to 8192 words. The only arithmetic operation was subtraction of a memory operand from the accumulator. The subtraction was carried out bit–serially with two's complement integers.

All storage elements, even the instruction register and program counter, were implemented by Wiliams tubes. In this section, we will redesign the MARK 1, starting from the original instruction set, but using the functional modules introduced earlier in this book; in particular, a bit-parallel arithmetic unit is used.

### 5.1.1 Instruction Set

The MARK 1 instruction format is listed in table 5.1. The three most significant bits specify the operation code, the next 13 bits specify a constant, and the 16 least significant bits are unused. Of course all 32 bits are used in the data paths.

| 31 | 30 | 29 | 28 | ... | 16 | 15 | ... | 0 |
|----|----|----|----|-----|----|----|-----|---|
| opcode | | | constant | | | unused | | |

Table 5.1: Instruction format of the MARK 1

We employ the register transfer language (RTL) described in appendix A to specify the semantics of the operations. The operations are summarized in table 5.2. The 13 bit constant "co" is always used for addressing the main memory; if a constant value is needed in a computation, the compiler must place it explicitly in the data section of the program code. Subtraction can be used to simulate addition $(a + b = a - (-b))$ and a left shift $(a << 1 = a + a)$. Bit-wise AND, OR and negation can be simulated by subtractions and the "Test" instruction and hence, the instruction set is powerful enough to compute standard bit operations. The "Stop" operation, which halts the machine, enabled the operating personal of the original MARK 1 to change main memory data via a console. Although this is, of course, not necessary for the architectural model, we keep this instruction to indicate the end of a calculation. During instruction fetch, the PC is always incremented. The "Test" operation increments the PC one more time, if the accumulator is negative. Since the values in the accumulator are stored in 2's complement format, the accumulator is negative, if and only if the most significant bit is 1.

| opcode | | | mnemonic | description |
|---|---|---|----------|-------------|
| 0 | 0 | 0 | Jmp | PC = M(co) |
| 0 | 0 | 1 | Jrel | PC = PC + M(co) |
| 0 | 1 | 0 | Load | ACC = −M(co); PC = PC+1 |
| 0 | 1 | 1 | Store | M(co) = ACC; PC = PC+1 |
| 1 | 0 | 0 | Sub | ACC = ACC − M(co); PC = PC+1 |
| 1 | 0 | 1 | Stop | Halt the machine |
| 1 | 1 | x | Test | If (ACC<0) PC = PC+2 <br> else PC = PC+1 |

Table 5.2: Instructions of the MARK 1

## 5.1.2  Data Paths

Figure 5.1 depicts the data paths of the MARK 1. The left side comprises the accumulator, the LMUX multiplexer and the subtractor which are 32 bit wide. The subtractor is implemented by a arithmetic unit (see figure 3.17) whose control input **sub** is permanently activated. In the arithmetic unit, a Carry Look-ahead Adder is employed. These components are used to execute Load, Store and Sub instructions.

The instruction register IR and the program counter PC are only affected by the other instructions and during fetch cycles. The instruction register IR is 16 bits wide and is connected to the higher 16 bits of the main memory. All other components are 13 bits wide.

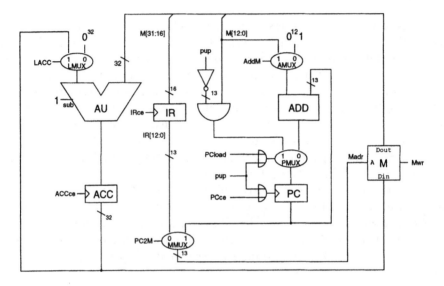

Figure 5.1: MARK 1 data paths

The multiplexer LMUX supports the execution of either Sub instructions or Load instructions. During execution of a Load instruction, the memory operand is subtracted from zero, and hence, the value $-M(co)$ is stored in the accumulator. During execution of a Sub instruction, the memory operand is subtracted from the accumulator, and hence, $ACC - M(co)$ is stored in the accumulator. The

| State | active control signals Csig | State | active control signals Csig |
|-------|------------------------------|-------|------------------------------|
| Fetch | PC2M, IRce, PCce | Store | Mwr |
| Decode | — | Sub | ACCce, LACC |
| Jump | PCload, PCce | Halt | — |
| Jrel | AddM, PCce | Test, taken | PCce |
| Load | ACCce | | |

Table 5.3: Active control signals for the MARK 1

main memory is addressed by the IR, except for fetch cycles in which the main
memory is addressed by the PC. The role of the MMUX is to select the address
either from the IR or from the PC.

The remaining modules implement the program counter. The PMUX selects
the new PC value from the adder in fetch cycles and in the execution of Jrel
and Test instructions. In the execution of a Jmp instruction and on powerup,
the PMUX selects the new PC value from the AND circuit. The sole purpose of
the AND circuit is to enable reset of the PC to zero during powerup. The adder
either increments the old PC value, or adds the value read from the memory to
the old PC value.

### 5.1.3   Control

Figure 5.2 depicts the control automaton of the MARK 1. Table 5.3 lists the
active control signals for each state. The Fetch state, which has the biggest
fanin, is encoded as state $z_0$. In the Decode state, which always follows Fetch,
no control signals are activated in the data paths, but the next state of the
control automaton is computed. The next state (see figure 4.2) depends on the
opcode, which is stored in the instruction register at the end of the Fetch cycle.
The fetch and decode states could be fused into one state, however, since the
opcode is only valid at the end of the fetch cycle, we prefer to separate these
states; see exercise 5.7 for further discussion of this issue.

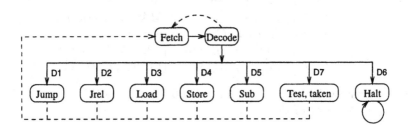

Figure 5.2: MARK 1 control automaton

The state fetch follows all states except for Halt and Decode. Once the Halt
state is reached, it can only be left by a new power up. The Decode state
is followed by a Fetch state, if the Opcode is TEST and the accumulator is
non-negative (see monomial of D7 in table 5.4). All other state transitions are
uniquely determined by the opcode. The condition monomials D1, . . . , D7 which
govern the state transitions are listed in table 5.4.

## 5.2   Evaluation

Evaluation of a design is carried out in four steps. First, we compute the cost of
the design. Then, the cycle time is calculated, and the run time of a benchmark

| Name | Monomial |
|------|----------|
| D1 | /IR[31] * /IR[30] * /IR[29] |
| D2 | /IR[31] * /IR[30] * IR[29] |
| D3 | /IR[31] * IR[30] * /IR[29] |
| D4 | /IR[31] * IR[30] * IR[29] |
| D5 | IR[31] * /IR[30] * /IR[29] |
| D6 | IR[31] * /IR[30] * IR[29] |
| D7 | IR[31] * IR[30] * ACC[31] |

Table 5.4: Monomials of the MARK 1

program is determined. Finally, based on these calculations, we can compute the quality of the architecture with regard to the specific benchmark.

## 5.2.1 Cost

The cost of the data paths is the sum of its component's cost. We read it off directly from figure 5.1. In accordance to the model, the cost of the main memory is left out. Hence, the cost of the data paths equals

$$
\begin{aligned}
C_{DP} = \quad & C_{mux}(32) + C_{AU}(32) && [LMUX, AU] \\
+ \quad & C_{ff}(32) + C_{ff}(16) && [ACC, IR] \\
+ \quad & C_{mux}(13) + C_{inv} && [MMUX, powerup] \\
+ \quad & 13 \cdot C_{and} + 2 \cdot C_{or} && [powerup] \\
+ \quad & C_{mux}(13) + C_{ADD}(13) && [AMUX, PCadder] \\
+ \quad & C_{ff}(13) + C_{mux}(13) && [PCreg., PMUX]
\end{aligned}
$$

The program which is listed in appendix B.5 evaluates this equation. Table 5.5 displays the result of that program.

| Component | Cost [g] | Component | Cost[g] |
|-----------|----------|-----------|---------|
| Accumulator | 256 | IR | 128 |
| LMUX | 96 | PC | 104 |
| Subtractor | 802 | AMUX | 39 |
| MMux | 39 | Adder | 328 |
| Powerup Logic | 31 | PMUX | 39 |
| **Total:** | **1862** | | |

Table 5.5: Cost of the MARK 1 data paths

The cost of the control logic is determined by the parameters listed in table 5.6. The number of input signals ($\sigma$) and output signals ($\gamma$) can be read off from table 5.3. The number of states (k) can be found in figure 5.2. The next parameter ($\zeta$) is the binary logarithm of k. The number of nontrivial monomials (#M), the maximal number of literals in a single monomial ($l_{max}$), and their cumulative length ($l_{sum}$) can be derived from the monomial table 5.4. The maximal ($\nu_{max}$) and cumulative ($\nu_{sum}$) signal activity is computed from table 5.3. Finally, the maximal ($\text{fanin}_{max}$) and cumulative fanin ($\text{fanin}_{sum}$) from figure 5.2 complete the parameter set.

| $\sigma$ | $\gamma$ | k | $\zeta$ | #M | $l_{max}$ | $l_{sum}$ | $\nu_{max}$ | $\nu_{sum}$ | $\text{fanin}_{max}$ | $\text{fanin}_{sum}$ |
|---|---|---|---|---|---|---|---|---|---|---|
| 4 | 8 | 9 | 4 | 7 | 3 | 21 | 4 | 12 | 2 | 9 |

Table 5.6: Parameters for the MARK 1 control circuit

The cost of the control automaton depends on the actual implementation (Mealy/Moore, unary/binary) and can be computed according to the formulae in chapter 4. Instead of evaluating the formulae by hand, we employ the evaluation program listed in appendix B.5 to accomplish this time consuming and error-prone task. The output of the program is shown in figure 5.3. The last row indicates, that the control logic contributes about 10 % to the total cost of the MARK 1 under the assumption that Carry Look-ahead adders are used.

```
cost                           Moore              Mealy
                          unary  binary      unary  binary
-----------------------------------------------------------
            data path:    1862    1862        1862    1862
              control:     197     229         259     315
-----------------------------------------------------------
                Mark1:    2059    2091        2121    2177
      control/Mark1 [%]:      9      10          12      14
```

Figure 5.3: Output of cost calculation

## 5.2.2  Cycle–Time

The cycle time is determined by the longest path from one storage element (register, RAM, ROM, register like constant or input signal) to a register or a RAM. The accumulated delay of the main memory address bus Madr is calculated as follows. Recall, that a control signal has an accumulated delay of $A_{CON}(Csig)$. Hence,

$$A_{Madr} = D_{mux}(13) + A_{CON}(Csig)$$

The accumulated delay for reading the main memory ($A_{Dout}$) and the delay of the adder ($A_{Add}$) are computed below.

$$A_{Dout} = D_{Mem} + \max\{A_{CON}(Csig), A_{Madr}\}$$
$$A_{Add} = D_{Add} + \max\{A_{CON}(Csig), A_{Dout}\}$$

The bound on cycle time imposed by the path through the accumulator is calculated as follows: One starts at the accumulator whose delay equals $\Delta = \delta + D_{ff}$, which models the flip flop delay $D_{ff}$ and the setup time $\delta$. Hence,

$$T_{ACC} = \Delta + D_{AU} + \max\{D_{mux}(32) + A_{CON}(Csig), \\ A_{Dout}\}$$

The remaining cycle time bounds imposed by paths through the IR, PC and Main Memory (in case of a write) are listed below. Note, that in the calculation of a memory write cycle only $\delta$ is used.

$$T_{IR} = \Delta + \max\{A_{CON}(Csig), A_{Dout}\}$$

$$T_{PC} = \Delta + D_{mux}(13) + \max\{A_{CON}(Csig) + D_{or}, \\ D_{AND} + \max\{A_{Dout}, \\ D_{Inv}\}, \\ A_{Add}\}$$

$$T_{Mem} = \delta + \max\{A_{CON}(Csig), \\ A_{Madr}\}$$

The final cycle time equals the maximum over all these cycle time bounds. As in the computation of the cost, these computations are performed by the program listed in appendix B.5. Its output, assuming a Carry Look-ahead Adder and a memory access time of $dmem = 20d$, is listed in figure 5.4.

| delay | Moore | | Mealy | |
| (dmem = 20) | unary | binary | unary | binary |
| --- | --- | --- | --- | --- |
| control signals: | 4 | 9 | 0 | 0 |
| memory read: | 26 | 31 | 22 | 22 |
| arithmetic unit: | 50 | 50 | 50 | 50 |
| memory write: | 26 | 31 | 22 | 22 |
| accumulator: | 81 | 86 | 77 | 77 |
| instr. register: | 31 | 36 | 27 | 27 |
| program counter: | 75 | 80 | 71 | 71 |
| data path: | 81 | 86 | 77 | 77 |
| control: | 21 | 24 | 23 | 24 |
| Mark1: | 81 | 86 | 77 | 77 |

Figure 5.4: Output of cycle time calculation

## 5.2.3　Run Time of a Benchmark

As mentioned in the introduction of this chapter, the original MARK 1 was built for testing the Wiliams tube memory. One of the programs, which was run on the MARK 1, is listed in figure 5.5 as a C–program. It computes the greatest factor of a given number using only iterative subtractions.

```
int gf(int x)              /* search greatest factor of x */
{ int f,h;                 /* factor f, temporary variable h */
  f=x;
  do {
    f--; h=x;              /* try the next smaller factor */
    do {
      h=h-f;              /* subtract the factor, until ...  */
      if(h==0) return (f);/* ... either h=0:  then, f is a factor */
    } while (h>0);         /* ... or h<0:  then, f is not a factor */
  } while(1);
}
```

Figure 5.5: Program "Greatest Factor" written in C

This program, which is specified in the high level language C, must be transformed into a machine program. We will do this translation by hand, because we have no compiler for the MARK 1. First, the program is rewritten in C notation (figure 5.6), using only assignments that map easily to MARK 1 instructions, and breaking down the do–loops into labels and gotos.

```
int gf(int x)           /* coding similar to MARK1 machine language*/
{ int f,h,t;            /* factor, helping variable, test condition */
  f=x;
outer:                  /* outer loop */
    f--; h=x;
inner:                  /* inner loop*/
    h=h-f;
    if(h<0)
      goto check;       /* check end condition for inner loop */
    goto inner;
check:
    t=-h-f;             /* t=-(h+f); undo last subtraction */
    if(t<0)             /* f is not a factor → continue */
      goto outer;
    return(f);
}
```

Figure 5.6: Restructured version of the program

Then, the C–program is translated into a MARK 1 machine program (see figure 5.7). The memory map is shown in the program header.

```
;;; Input :    x in M[0]
;;; Output:    f in M[1]
;;; Variables: h in M[2], t in ACC
begin:    ACC  := - M(0)    ;
          M(0) := ACC       ; M(0) = -x (gives better code
          ACC  := - M(0)    ;           for initializing h)
          M(1) := ACC       ; f := M(1) = x
outer:    ACC  := - M(1)    ; outer loop
          M(1) := ACC       ;
          ACC  := - M(1)    ; load f (negated twice by load)
          ACC  := ACC - 1   ; f--
          M(1) := ACC       ; store f to memory
          ACC  := - M(0)    ; init h = -(-x)
inner:    ACC  := ACC - M(1); h = h - f
          If (ACC<0) check  ; test loop condition
          JMP  inner        ; next iteration of inner loop
check:    M(2) := ACC       ; store h
          ACC  := - M(1)    ; load -f and ..
          ACC  := ACC - M(2); .. subtract h
          If (ACC<0) nextout;
          Stop              ; ready
nextout:  JMP  outer
```

Figure 5.7: Machine language version

**Program analysis** On the original MARK 1, the program was run with a input value of $2^{18}$. Since the greatest factor of this number is $2^{17}$, the outer loop has to be executed $2^{17}$ times. We therefore define the benchmark $Be$ to be *one* pass of the outer loop, that is, checking if $f$ is a factor of $x$. For $x = 2^{18}$ and $f \in [2^{17}, 2^{18}]$, the inner loop is performed twice, and the check is performed once.

| Instruction | outer | inner (2×) | check | sum | $W_I$ | $CPI_I$ |
|---|---|---|---|---|---|---|
| Jmp | — | 2 | — | 2 | 12,50 | 3 |
| Jrel | — | — | — | — | 0,00 | 3 |
| Load | 3 | — | 1 | 4 | 25,00 | 3 |
| Store | 2 | — | 1 | 3 | 19,75 | 3 |
| Sub | 1 | 2 | 1 | 4 | 25,00 | 3 |
| Test (taken) | — | 1 | 1 | 2 | 12,50 | 3 |
| Test (not taken) | — | 1 | — | 1 | 6,25 | 2 |
| | | | | | CPI: 2.94 | |

Table 5.7: Instruction mix of the MARK 1 benchmark

The CPI of the benchmark is computed in table 5.7. We count the number of instructions in each part of the benchmark. Then, we sum the total number of instructions in an execution of the benchmark, and compute the relative frequencies $W_I$ of the instructions. Each instruction is executed in three clock cycles (Fetch, Decode, and the specific execution cycle), except for a Test, which is not taken. One pass of the outer loop contains $IC = 16$ instructions and lasts $IC \cdot CPI = 47$ cycles.

## 5.2.4  Quality

Now, we can compute the quality of the different MARK 1 versions. The quality itself does not give much evidence about one single design, but the changes in the quality ratio between different designs help comparing them.

| Variant | cost [g] | $t_c$ [d] | T($Be$) [d] |
|---|---|---|---|
| Moore, unary | 2055 | 81 | 3807 |
| Moore, binary | 2087 | 86 | 4042 |
| Mealy, unary | 2117 | 77 | 3619 |
| Mealy, binary | 2173 | 77 | 3619 |

Table 5.8: Cost and time for MARK 1 variants

Table 5.8 summarizes cost, cycle time $t_c$ and benchmark time T($Be$), assuming a Carry Look-ahead Adder and a memory access time of $dmem = 20d$. It turns out, that the binary versions are more expensive for both implementation types; for the Moore type, binary coding is even slower. This is in accordance with table 4.2.

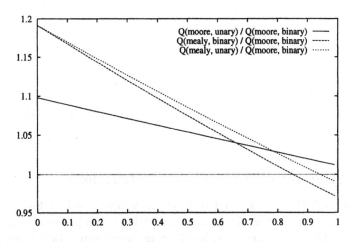

Figure 5.8: Quality as a function of $q$

Figure 5.8 depicts the quality ratio as a function of the quality parameter, normalized to the "Moore automaton/binary coding" version. Here again, one clearly sees, that the unary versions are superior to the binary versions. The best implementation for a quality parameter $< 0.8$ is "Mealy, unary". Only if the cost of a design is much more important than the performance, the "Moore, unary" version would be chosen.

## 5.3   Trade–Off Analysis

In this section, we will discuss some design variations which change cost and/or performance. A few of these modifications diminish cost, but do not impair performance, thus, it is clear which design to choose. Other modifications involve a cost–performance tradeoff. In those cases, a detailed trade–off analysis will show, which design is better in which situations.

### 5.3.1   Using Special Purpose Components

The components LMUX and AMUX are multiplexers with one constant value input. Figure 5.9 shows a realization of LMUX and AMUX, which is more cost–effective than a general purpose multiplexer. This realization saves $44g$, which equals 2% of the total MARK 1 cost. Since the cycle time remains unchanged, this modification improves quality by 0.4% to 1.1% for a quality parameter $q \in [0.2, 0.5]$.

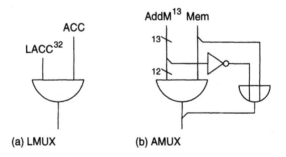

Figure 5.9: Realization of multiplexers: (a) LMUX, (b) AMUX

### 5.3.2   Eliminating Superfluous Instructions

According to table 5.7, the Jrel instruction is never used in the benchmark. Moreover, a compiler can always simulate it by a Jump instruction. So, the AMUX and the adder can be replaced by an incrementer, that saves 321 gate equivalents (15%). Furthermore, the control logic gets simpler (see table 5.9), reducing the cost by about $30g$.

| | $\sigma$ | $\gamma$ | k | $\zeta$ | #M | $l_{max}$ | $l_{sum}$ | $\nu_{max}$ | $\nu_{sum}$ | fanin$_{max}$ | fanin$_{sum}$ |
|---|---|---|---|---|---|---|---|---|---|---|---|
| old | 4 | 8 | 9 | 4 | 7 | 3 | 21 | 4 | 12 | 2 | 9 |
| new | 4 | 7 | 8 | 3 | 6 | 3 | 18 | 3 | 10 | 2 | 8 |

Table 5.9: Parameters for the control circuit

## 5.3.3  Conditional Jump Optimization

The current instruction set makes conditional jumps in the benchmark rather cumbersome. The reason is that the Jrel conditional jump only enables skipping a single instruction. This requires an additional backward jump for constructing program loops. An example for this behavior is the last test in **check**:

```
         If (ACC<0) nextout;
         Stop                 ; ready
nextout: JMP   outer
```

We modify the Test instruction so that the jump target address is encoded in the instruction. One possibility to do this, is to connect the left PMUX input directly to IR[28:16]. Figure 5.10 depicts the resulting data paths of a machine, denoted by M1_0.

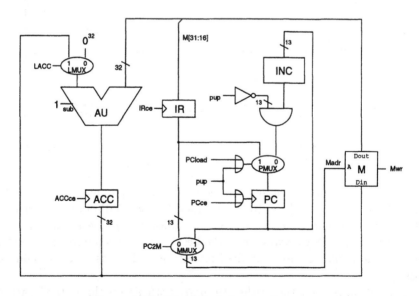

Figure 5.10: The M1_0 data paths

### 5.3.4 Different Adder Implementations

The choice of the adder changes the run time of the benchmark, and the cost of the design. Table 5.10 lists the parameters for the different implementations of the arithmetic unit.

| Arithmetic Unit | $C_{AU}$ [g] | $D_{AU}$ [d] | $t_c$ [d] |
|---|---|---|---|
| Ripple Carry (RCA) | 584 | 134 | 161 |
| Carry look-ahead (CLA) | 802 | 50 | 77 |
| Conditional sum (CSA) | 1285 | 20 | 47 |

Table 5.10: Cost and time for AU variants

The quality parameter models the relative impact of the changes in cost and time. Figure 5.11 depicts the quality ratio of different adder implementations as a function of the quality parameter $q$ assuming a memory delay of $dmem = 20d$.

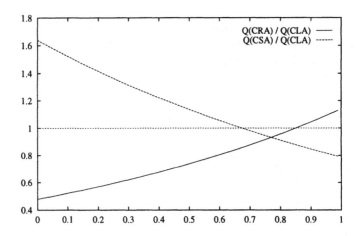

Figure 5.11: Quality for $dmem = 20d$ depending on the quality parameter

The CSA version is the best choice for a quality parameter $q < 0.65$; CLA is best for $q \in [0.65, 0.85]$ and RCA wins for $q > 0.85$. Within the parameter range $[0.2, 0.5]$, which we consider reasonable values for $q$ (see discussion following definition 2.2), CSA always yields the best result. This unintuitive effect is due to the fact, that we use unlimited fanout here. Exercise 5.2 will address this issue.

The memory and the arithmetic unit, both lie on the critical path of the M1_0. Thus, the quality of the whole architecture does not only depend on the adder implementation, as suggested in the last figure, but it also depends on the memory access time. Figure 5.12 depicts the equal quality parameters of M1_0

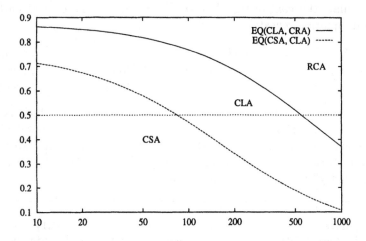

Figure 5.12: Equal quality parameter depending on *dmem*

design variants with different adder types, as a function of the memory access time *dmem*. The curves partition the diagram area into three regions, each of them defining the best design variant for a specific pair (*dmem*, *q*).

## 5.3.5   Introducing a Separate Memory Cycle

We have seen, that the critical path goes through main memory and the subtractor. Splitting up this path therefore reduces the cycle time. We place a register MDR (*memory data register*) between main memory and the subtractor and change the control automaton as sketched in figure 5.13. Now, Load and Sub will need 4 cycles instead of 3, but this loss should be compensated by the shorter cycle time. We call the new machine M1_1.

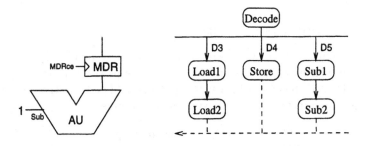

Figure 5.13: Changes leading to design M1_1

If the adder is implemented by a Conditional Sum Adder, then the PC incrementer (ripple carry style) and the powerup logic lie on the critical path.

In particular, the control signals $Csig$ do not lie on the critical path, as the following equation reveals:

$$T_{PC} = \Delta + D_{mux}(13) + \max\{A_{CON}(Csig), \quad ; \text{PCLoad}$$
$$D_{And} + \max \{ D_{Inv}, \quad ; \text{powerup}$$
$$D_{Inc} \} \} \quad ; \text{incrementer}$$

Figure 5.14 lists the output of the evaluation program which is listed in appendix B.5.2. The cycle time is limited to $36d$ for the Moore Automaton with binary coding and $33d$ for the other designs. Since the Moore implementation of the control automaton is cheaper than the Mealy implementation, "Moore/unary" is the best design variant.

| delay            | Moore | | Mealy | |
| (dmem = 20)      | unary | binary | unary | binary |
| --- | --- | --- | --- | --- |
| control signals: | 4 | 9 | 0 | 0 |
| memory read:     | 26 | 31 | 22 | 22 |
| arithmetic unit: | 20 | 20 | 20 | 20 |
| memory write:    | 27 | 32 | 23 | 23 |
| memory data reg.: | 31 | 36 | 27 | 27 |
| accumulator:     | 31 | 36 | 27 | 27 |
| instr. register: | 31 | 36 | 27 | 27 |
| program counter: | 33 | 33 | 33 | 33 |
| data path:       | 33 | 36 | 33 | 33 |
| control:         | 23 | 24 | 25 | 25 |
| M1_1:            | 33 | 36 | 33 | 33 |

Figure 5.14: cycle time calculation for M1_1

## 5.3.6  Adding a New Instruction

Besides reducing cost or cycle time, one can improve quality by spending less cycles in the benchmark; i.e. using less, but more powerful instructions. In the "Greatest Factor" benchmark, a new instruction "Load positive" (Ldpos) can replace the three instructions following the label **outer**:

```
outer:   ACC  := - M(1)    ; outer loop
         M(1) := ACC       ;
         ACC  := - M(1)    ; load f (negated twice by load)
```

The new instruction can be encoded in place of the old Jrel instruction. The AU control input Sub is not longer tied to 1, but receives a control signal AUsub, which gets activated for Load and Sub instructions. Applying these changes to the M1_1 results in a design M1_2, whose control circuit parameters are shown in table 5.11. The cost increases by 24 gates. This design modification saves 7 cycles (15%) of the benchmark.

| $\sigma$ | $\gamma$ | k | $\zeta$ | #M | $l_{max}$ | $l_{sum}$ | $\nu_{max}$ | $\nu_{sum}$ | fanin$_{max}$ | fanin$_{sum}$ |
|---|---|---|---|---|---|---|---|---|---|---|
| 4 | 8 | 8 | 3 | 7 | 3 | 21 | 3 | 10 | 2 | 7 |

Table 5.11: Parameters for the M1_2 control circuit

### 5.3.7  Summary

Figure 5.15 depicts the quality of all improved versions at a glance. While M1_0 saves cost with no performance decreased, M1_1 further increases performance by shortening the cycle time at the expence of cost. In addition, the M1_2 design decreases the number of cycles needed in the benchmark and improves quality even further. We conclude that the best design is the M1_2, since we consider the range [0.2, 0.5] of quality parameter to be the interesting range.

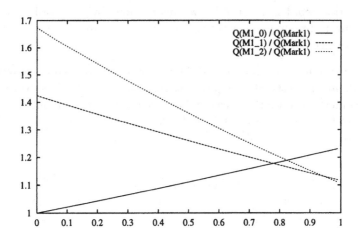

Figure 5.15: Quality ration of M1_0, M1_1 and M1_2 vs. MARK 1 as a function of the quality parameter $q$

## 5.4 Exercises

5.1 Sketch the control automaton for the M1_0 and verify the parameters given in table 5.9.

5.2 Use the fanout restricted fast carry look-ahead adder and conditional sum adders from exercises in chapter 3.9. Repeat the trade–off analysis shown in figure 5.11 and 5.12 using these components.

5.3 Construct a conditional sum incrementer and use it in design M1_1. Evaluate and compare the quality of the new design with the M1_1 design. Does the Mealy version get better than the Moore version?

5.4 In the M1_2 design, the actions taken in the Decode state can be extended to read M(co) from Memory, and store it in the memory data register. Then, the Load and Sub cycles do not have to be split, resulting in a lower CPI value. Specify and evaluate this design variant.

5.5 In the instruction word, 16 bits are unused. A compiler could encode the "next PC" in this space. Change the design in such a way, that the incrementer becomes obsolete. By how much is quality improved?

5.6 An alternative approach that takes advantage of the free 16 bits of a instruction word is to encode two instructions in one word. Change the hardware for supporting this feature and compare the results with those of exercise 5.5.

5.7 An approach for avoiding the Decode state in the above designs, is to evaluate the instruction register *input* in state Fetch. The drawback with this method is, that the accumulated delay $A_{In}$ of the input signals of the control automaton is as high as the memory access time *dmem*. Then, the control lies on the critical path.

1. Calculate the cycle time of this design.

2. In the Mealy implementation, the control outputs Csig are stored in a register. Can the above trick be used for Mealy type control automata, too?

# Chapter 6

# Design of the DLX Machine

The remainder of this book deals with the implementation of a non-pipelined, fixed-point DLX (Deluxe) machine and the evaluation of some design changes. Hennessy and Patterson introduced the DLX architecture in [HP90] and sketched several implementations. Our architecture model requires a somewhat more detailed specification of the DLX hardware. The extra details concern glue logic which depends directly on the coding of the instructions. Technically, this material is in our view reasonably systematic and quite valuable, because it makes the impact of good or bad coding visible.

For the sake of simplicity, we first ignore interrupt handling. Chapter 8 then describes interrupts and their implementation in all detail.

## 6.1 Instruction Set Architecture

DLX is a RISC architecture with only three instruction formats. It uses 32 general purpose registers, each 32 bits wide. Register R0 always has the value 0.

Load and store operations move data between the general purpose registers and the memory. There is a single addressing mode: the effective memory address *ea* is the sum of a register and an immediate constant. Except for shifts, immediate constants are *always* sign-extended to 32-bits.

The memory is byte addressable and performs byte, half-word or word accesses. All instructions are coded in four bytes. In memory, data and instructions must be aligned in the following way: Half words must start at even byte addresses. Words and instructions must start at addresses divisible by 4. These addresses are called word boundaries.

The DLX architecture also has a few special purpose registers mainly used for handling interrupts. Special move instructions transfer data between general and special purpose registers.

## 6.1.1   Instruction Formats

All three instruction formats (figure 6.1) have a 6-bit primary opcode and specify up to three explicit operands. The I-type (Immediate) format specifies two
registers and a 16-bit constant. That is the standard layout for instructions with
an immediate operand. The J-type (Jump) format is used for control instructions. They require no explicit register operand and profit from the larger 26-bit
immediate operand. The third format, R-type (Register) format, provides an
additional 6-bit opcode (*function*). The remaining 20 bits specify three general
purpose registers and a field $SA$ which specifies a 5-bit constant or a special
purpose register. A 5-bit constant, for example, is sufficient as shift amount.

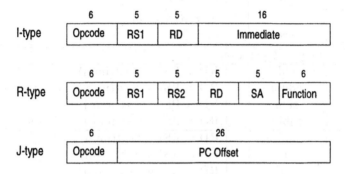

Figure 6.1: The three instruction formats of the DLX design. RS1 and RS2 are
source registers; RD is the destination register. SA specifies a special purpose
register or an immediate shift amount. Function is an additional 6-bit opcode.

## 6.1.2   Instruction Set Coding

Since the DLX description in [HP90] does not specify the coding of the instruction set, we adapt the coding of the MIPS R2000 machine ([HP93, KH92]) to the
DLX instruction set. Tables 6.1 through 6.3 specify the instruction set and list
the coding; the prefix "hx" indicates that the number is represented as hexadecimal. The effects of the instructions are specified in a register transfer language
(see appendix A).

## 6.1.3   Comments

The branch target computation and the coding of the register operands may
seem at first sight somewhat unusual.

### Branch Target Computation

Control operations compute the branch target relative to the address of the instruction sequentially following the control instruction. Thus, the branch target

| IR[31 : 26] | Mnemonic | d | Effect |
|---|---|---|---|
| **Data Transfer** | | | |
| hx20 | lb | 1 | RD = Sext(m) |
| hx21 | lh | 2 | RD = Sext(m) |
| hx23 | lw | 4 | RD = m |
| hx24 | lbu | 1 | RD = $0^{24}$m |
| hx25 | lhu | 2 | RD = $0^{16}$m |
| hx28 | sb | 1 | m = RD[7 : 0] |
| hx29 | sh | 2 | m = RD[15 : 0] |
| hx2b | sw | 4 | m = RD |
| **Arithmetic, Logical Operation** | | | |
| hx08 | addi | | RD = RS1 + Sext(imm) |
| hx09 | addiu | | RD = RS1 + Sext(imm)    no overflow |
| hx0a | subi | | RD = RS1 - Sext(imm) |
| hx0b | subiu | | RD = RS1 - Sext(imm)    no overflow |
| hx0c | andi | | RD = RS1 $\land$ Sext(imm) |
| hx0d | ori | | RD = RS1 $\lor$ Sext(imm) |
| hx0e | xori | | RD = RS1 $\otimes$ Sext(imm) |
| hx0f | lhi | | RD = $imm0^{16}$ |
| **Test Set Operation** | | | |
| hx19 | sgri | | RD = (RS1 > Sext(imm)  ?  1 : 0); |
| hx1a | seqi | | RD = (RS1 = Sext(imm)  ?  1 : 0); |
| hx1b | sgei | | RD = (RS1 $\geq$ Sext(imm)  ?  1 : 0); |
| hx1c | slsi | | RD = (RS1 < Sext(imm)  ?  1 : 0); |
| hx1d | snei | | RD = (RS1 $\neq$ Sext(imm)  ?  1 : 0); |
| hx1e | slei | | RD = (RS1 $\leq$ Sext(imm)  ?  1 : 0); |
| **Control Operation** | | | |
| hx04 | beqz | | PC = PC + 4 + (RS1 = 0  ?  Sext(imm) : 0) |
| hx05 | bnez | | PC = PC + 4 + (RS1 $\neq$ 0  ?  Sext(imm) : 0) |
| hx16 | jr | | PC = RS1 |
| hx17 | jalr | | R31 = PC + 4;    PC = RS1 |

Table 6.1: I-type instruction layout. All instructions except the control instructions also increment the PC, $PC$ += $4$. *Sext(a)* is the sign-extended version of $a$. The effective address of memory accesses equals $ea = RS1 + Sext(imm)$, where $imm$ is the 16-bit intermediate. The width of the memory access in bytes is indicated by $d$. Thus, the memory operand equals $m = M[ea+d-1], \cdots, M[ea]$.

| IR[31 : 26] | IR[5 : 0] | Mnemonic | Effect |
|---|---|---|---|
| Shift Operation | | | |
| hx00 | hx00 | slli | RD = RS1 << SA |
| hx00 | hx02 | srli | RD = RS1 >> SA |
| hx00 | hx03 | srai | RD = RS1 $>>_{\text{arith}}$ SA |
| hx00 | hx04 | sll | RD = RS1 << RS2[4 : 0] |
| hx00 | hx06 | srl | RD = RS1 >> RS2[4 : 0] |
| hx00 | hx07 | sra | RD = RS1 $>>_{\text{arith}}$ RS2[4 : 0] |
| Arithmetic, Logical Operation | | | |
| hx00 | hx20 | add | RD = RS1 + RS2 |
| hx00 | hx21 | addu | RD = RS1 + RS2    no overflow |
| hx00 | hx22 | sub | RD = RS1 - RS2 |
| hx00 | hx23 | subu | RD = RS1 - RS2    no overflow |
| hx00 | hx24 | and | RD = RS1 ∧ RS2 |
| hx00 | hx25 | or | RD = RS1 ∨ RS2 |
| hx00 | hx26 | xor | RD = RS1 ⊗ RS2 |
| Test Set Operation | | | |
| hx00 | hx29 | sgr | RD = (RS1 > RS2 ? 1 : 0); |
| hx00 | hx2a | seq | RD = (RS1 = RS2 ? 1 : 0); |
| hx00 | hx2b | sge | RD = (RS1 ≥ RS2 ? 1 : 0); |
| hx00 | hx2c | sls | RD = (RS1 < RS2 ? 1 : 0); |
| hx00 | hx2d | sne | RD = (RS1 ≠ RS2 ? 1 : 0); |
| hx00 | hx2e | sle | RD = (RS1 ≤ RS2 ? 1 : 0); |
| Data Transfer | | | |
| hx00 | hx10 | movs2i | RD = SA    (see Chapter 8) |
| hx00 | hx11 | movi2s | SA = RS1    (see Chapter 8) |

Table 6.2: R-type instruction layout. All instructions execute $PC \mathrel{+}= 4$. SA is a special purpose register respectively the 5-bit immediate shift amount specified by the bits IR[10 : 6]. The two move operations are necessary to support interrupt handling.

| IR[31 : 26] | Mnemonic | Effect |
|---|---|---|
| Control | | |
| hx02 | j | PC = PC + 4 + Sext(imm) |
| hx03 | jal | R31 = PC + 4;    PC = PC + 4 + Sext(imm) |
| hx3e | trap | cause trap interrupt    (see Chapter 8) |
| hx3f | rfe | return from exception    (see Chapter 8) |

Table 6.3: J-type instruction layout. Sext(imm) is the sign-extended version of the 26-bit immediate called PC Offset.

equals $PC + 4 + \text{offset}$ instead of $PC + \text{offset}$. This is not a DLX specific phenomenon. It also occurs in other RISC processors, like MIPS R2000 [KH92] and Intel i860 [Int89], and there is a good reason, as we will see in section 6.3.1.

### Operand Coding

The position of the 2 to 3 register operand fields seems somewhat unusual too. Field RS1 specifies the first source register and RS2 the second one. But, depending on the instruction format, the address of the destination register $ARw$ is specified by different fields of the instruction register $IR$ or by a constant, namely

$$ARw = \begin{cases} IR[20:16] & \text{for I-type instructions} \\ IR[15:11] & \text{for R-type instructions} \\ 11111 & \text{for instructions jal and jalr} \end{cases}$$

That requires two multiplexers with select signals $Itype$ and $Jlink$ (jump and link) for selecting the proper write address of the register file (figure 6.2).

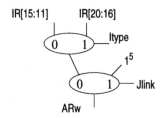

Figure 6.2: Select circuit of the destination register address $ARw$

It is tempting to switch the positions of the RS2 and RD fields in the R-type format. However, we would not get rid of all the collisions: store operations have type I and specify the memory address in the RD field. This destination must be *read* from the register file. Thus, switching the fields would complicate the selection of the second read address of the register file.

## 6.2  The Data Paths

Having introduced the DLX instruction set in the previous section, we now present the data paths of a non-pipelined implementation. Each instruction is executed in several cycles like: instruction fetch, instruction decode, operand fetch, execute, memory access and write back. Depending on the instruction, one or the other cycle is dropped.

The block diagram (figure 6.3) is almost literally copied from [HP90]. It shows the essential parts of the data paths and how they are connected to the memory system. However, we replaced the usual components like PC, ALU, shifter, etc., by what we call *environments* of these components. The environment of a

component consists of the component together with glue logic, which adapts the
component to the particular instruction set realized. Construction of this glue
logic is an essential and nontrivial part of hardware design, but it is hardly ever
covered in textbooks. A rare exception is [HP93].

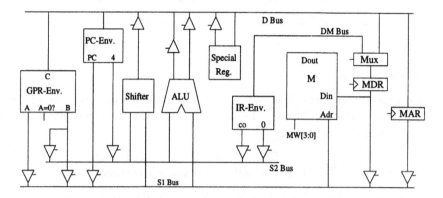

Figure 6.3: Block diagram of the DLX data paths and memory system

In the following subsections we will elaborate the various environments in the
following order: arithmetic logic unit (ALU), shifter (SH), instruction register
(IR), general purpose registers (GPR), program counter (PC), and memory (M).
The control unit generates the control signals for all these environments. We
use the following naming convention, to simplify matters.

**Naming of Control Signals**

The most common control signals are clocks and output enable signal of tristate
drivers. We use the following two rules to name those control signals:

First, a register name with the suffix *ce* indicates the clock signal of the
corresponding register. Second, enable signals of tristate drivers have the suffix
*doe*. The prefix indicates, where the data are coming from and where they are
going to. So, *AS1doe* is the enable signal of the driver between the A register
and the S1 bus.

## 6.2.1   The ALU Environment

The arithmetic logic unit (ALU) of the DLX machine should perform the following operations: addition, subtraction and comparison of two two's complement
numbers, three logical operations, namely the bitwise *and, or* and *xor*, and com-
posing a 32-bit integer constant by shifting a 16-bit constant 16 bits to the left.

The ALU environment therefore comprises the arithmetic unit AU(32) of
section 3.7, a comparator, a logical unit, and some glue logic. The AU is based
on the carry look-ahead adder CLA (section 3.7.3), because that is the adder
type most commonly used.

| Condition | | $a > b$ | $a = b$ | $a \geq b$ | $a < b$ | $a \neq b$ | $a \leq b$ |
|---|---|---|---|---|---|---|---|
| $f_2$ | $<$ | 0 | 0 | 0 | 1 | 1 | 1 |
| $f_1$ | $=$ | 0 | 1 | 1 | 0 | 0 | 1 |
| $f_0$ | $>$ | 1 | 0 | 1 | 0 | 1 | 0 |

Table 6.4: Specification of the test condition

**The Comparator**

Comparisons are performed during test set operations, which may be of type R or of type I. The condition to be tested is coded by three bits f[2:0] of the instruction register IR, namely

$$f[2:0] = \begin{cases} \text{IR}[2:0] & \text{for R-type instructions} \\ \text{IR}[28:26] & \text{for I-type instructions} \end{cases}$$

In either case, table 6.4 indicates the coding of the 6 possible conditions. This is a very common way to code conditions. The obvious implementation proceeds in 2 steps:

1. compute signals $l, e, g$ (less, equal, greater) with

$$\begin{aligned} l = 1 &\quad\leftrightarrow\quad a < b \quad\leftrightarrow\quad a - b < 0 \\ e = 1 &\quad\leftrightarrow\quad a = b \quad\leftrightarrow\quad a - b = 0 \\ g = 1 &\quad\leftrightarrow\quad a > b \quad\leftrightarrow\quad a - b > 0 \end{aligned}$$

2. the condition code is

$$zc = f_2 \wedge l \ \vee \ f_1 \wedge e \ \vee \ f_0 \wedge g.$$

Figure 6.4 depicts a realization along these lines. Using the arithmetic unit $AU$ as a subtractor, signal $l$ directly corresponds to the output flag $neg$ of $AU$. Signal $e$ is generated from the sum bits $s[31:0]$ by a 32-bit zero tester, and signal $g$ is computed as

$$g = \overline{e} \wedge \overline{l}$$

The 32-bit zero tester requires far more time than the delay of an inverter or of an AND or OR gate. Based on the associativity of the functions AND and OR, we generate signal $zc$ as:

$$\begin{aligned} zc &= f_2 l \ \vee \ f_1 e \ \vee \ f_0 (\overline{l}\,\overline{e}) \\ &= (f_2 l \ \vee \ f_1 e) \ \vee \ (f_0 \overline{l})\overline{e}. \end{aligned}$$

Thus, the additional cost and delay of the comparator Comp are

$$\begin{aligned} C_{Comp}(n) &= C_{zero}(n) + 2 \cdot C_{inv} + 4 \cdot C_{and} + 2 \cdot C_{or} \\ D_{Comp}(n) &= \max\{D_{inv} + D_{and}, \ D_{zero}(n) + D_{or}, \\ &\qquad D_{zero}(n) + D_{inv}\} + D_{and} + D_{or}. \end{aligned}$$

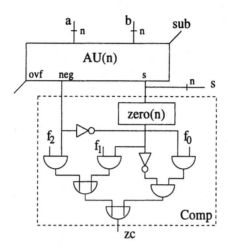

Figure 6.4: Arithmetic unit supplemented by the comparator circuit

### The Logic Unit

Arithmetic and logic operations are performed during instructions of type R or of type I. The operation is coded by the same three bits f[2:0] of the instruction register IR as the condition in test set operations. Table 6.5 indicates, how the operations are coded.

The circuit of figure 6.5 uses n-bit AND, OR and XOR gates and a few multiplexers to implement the logic unit and to connect it to the arithmetic unit. The cost of the ALU circuit therefore adds up to

$$C_{ALU}(n) = C_{AU}(n) + C_{Comp}(n) + n \cdot (C_{and} + C_{or} + C_{xor}) + 4 \cdot C_{mux}(n).$$

Later analysis will show that the compare logic of the ALU lies on the time critical path. We therefore differentiate the delays of the three outputs $zc$, $zr$ and $ovf$:

$$\begin{aligned}
D_{ALU}(zr;n) &= \max\{D_{AU}(n) + D_{mux}(n), \\
&\qquad \max\{D_{and}, D_{or}, D_{xor}\} + 3 \cdot D_{mux}(n)\} \\
D_{ALU}(zc;n) &= D_{AU}(n) + D_{Comp}(n) \\
D_{ALU}(ovf;n) &= D_{AU}(n).
\end{aligned}$$

### Connecting the ALU Environment to the Data Paths

Figure 6.6 depicts how to connect the ALU environment with the data paths. The three least significant bits of the opcodes (IR[28:26] or IR[2:0]) define the ALU function. The signal *Itype*, generated by the control, selects between the first and the second opcode. Some operations implicitly require an addition.

| | a+b | a-b | $a \wedge b$ | $a \vee b$ | $a \otimes b$ | $b[15:0]0^{n-16}$ |
|---|---|---|---|---|---|---|
| $f_2$ | 0 | 0 | 1 | 1 | 1 | 1 |
| $f_1$ | 0 | 1 | 0 | 0 | 1 | 1 |
| $f_0$ | x | x | 0 | 1 | 0 | 1 |

Table 6.5: Coding of the arithmetic/logical ALU operations

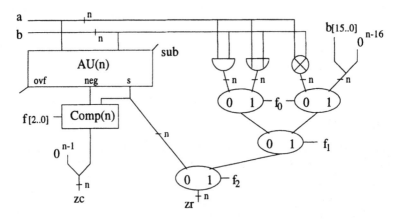

Figure 6.5: n-bit ALU circuit

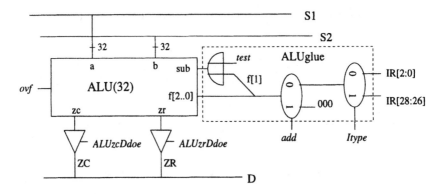

Figure 6.6: Implementation of the ALU environment

Control signal *add* enforces such an addition. The arithmetic unit AU inside the ALU performs subtractions when specified in the opcode ($f[1] = 1$) or during test set instructions. In the latter case, control signal *test* triggers the subtraction, and the flags $f[2 : 0]$ select the test condition. Note that on test set instructions, the output of the arithmetic unit may not be forwarded to the ALU output $zr$. An implicit subtraction (not just a test) can only be enforced by the combination of the *add* and *test* signal. The control unit further provides the enable signals $ALUzcDdoe$ and $ALUzrDdoe$ of the two tristate drivers.

The cost of the whole environment is

$$C_{ALUenv} = C_{ALU}(32) + 2 \cdot C_{driv}(32) + 2 \cdot C_{mux}(3) + C_{or}.$$

Some paths through $AU$ and zero tester have a large delay. So, they are candidates for the critical path in this design. Therefore, we model the delays of this environment in more detail.

The ALU has three outputs and two types of inputs, namely the data inputs S1 and S2, and control inputs. The instruction word IR and some signals $Csig$ from the control unit serve as control inputs. For each combination of inputs and outputs of the ALU environment, we specify a separate delay. The glue logic of the ALU environment generates the signals *sub* and $f[2 : 0]$. That takes time

$$D_{ALUglue} = 2 \cdot D_{mux}(3) + D_{or}.$$

The delays of the six different paths through the ALU environment can then be expressed as:

$$
\begin{aligned}
D_{ALUenv}(S1, S2; ZC) &= D_{ALU}(zc; 32) + D_{driv} \\
D_{ALUenv}(S1, S2; ZR) &= D_{ALU}(zr; 32) + D_{driv} \\
D_{ALUenv}(S1, S2; ovf) &= D_{ALU}(ovf; 32) \\[6pt]
D_{ALUenv}(Csig; ZC) &= D_{ALUglue} + D_{ALU}(zc; 32) + D_{driv} \\
D_{ALUenv}(Csig; ZR) &= D_{ALUglue} + D_{ALU}(zr; 32) + D_{driv} \\
D_{ALUenv}(Csig; ovf) &= D_{ALUglue} + D_{ALU}(ovf; 32).
\end{aligned}
$$

## 6.2.2   The Shifter Environment

The shifter environment requires very careful engineering. Shifts are not only performed during explicit shift operations (R-type) but also during load and store operations. We call shifts of the second kind *special shifts*. The environment performs three types of explicit shifts, namely logical left shifts, logical right shifts and arithmetical right shifts.

**Definition 6.1 (Explicit Shifts)** *Let $n = 2^m$. An n-bit logical left shift, logical right shift and arithmetical right shift compute the functions ll, rl, ra :*

| IR[1:0] | mnemonic | fill | mask |
|---------|----------|------|------|
| 00 | slli, sll | 0 | $0^{32-dist}1^{dist}$ |
| 10 | srli, srl | 0 | $1^{dist}0^{32-dist}$ |
| 11 | srai, sra | $a_{31}$ | $1^{dist}0^{32-dist}$ |

Table 6.6: Fill bit and mask of explicit shifts

$$\{0,1\}^{n+m} \to \{0,1\}^n, \ (a_{n-1},\ldots,a_0,b_{m-1},\ldots,b_0) \mapsto (c_{n-1},\ldots,c_0), \ with$$

$$(c_{n-1},\ldots,c_0) = \begin{cases} (a_{n-\langle b\rangle-1},\ldots,a_0,0^{\langle b\rangle}) & , \ for \ ll \\ (0^{\langle b\rangle},a_{n-1},\ldots,a_{\langle b\rangle}) & , \ for \ rl \\ (a_{n-1}^{\langle b\rangle},a_{n-1},\ldots,a_{\langle b\rangle}) & , \ for \ ra \ . \end{cases}$$

The core of the shift environment is the 32-bit cyclic left shifter CLS(32) of section 3.8. Shift operations are characterized by four parameters:

1. the shift distance $dist \in \{0,\ldots,31\}$.

2. the shift direction $right \in \{0,1\}$,

3. a bit $fill \in \{0,1\}$, and

4. a string $mask \in \{0,1\}^{32}$ indicating which output bits of the cyclic shifter have to be replaced by the fill bit.

The $a$-input of the shifter always specifies the operand which has to be shifted. For explicit shifts, the distance $dist$ is taken from the $b$-input of the shifter. The bits IR[1:0] of the instruction register determine the mask, and fill bits according to table 6.6.

For special shifts, distance, mask and fill bit depend on

1. the type of memory access (load or store), specified by IR[29]

2. the bit IR[28] which indicates whether the load is signed or unsigned

3. the width of the memory access specified by IR[27:26]

4. the last two bits MAR[1:0] of the memory address register MAR.

We always perform special shifts as *left* shifts and specify the distance accordingly. For the shift distance, it does not matter whether a load is signed or unsigned. Thus, the distance determined by table 6.7 is independent of IR[28].

A nonzero fill bit is only needed for the sign extension in instructions lb and lh. Thus, the fill bit can be determined according to table 6.8.

For store instructions, the bank write signals of main memory extract the proper data bytes. Thus, nontrivial masks are only needed for instructions lb, lbu, lh and lhu. The mask is applied to the output of the shifter, so the memory

| IR[29,27,26] | mnemonic | MAR[1:0] | dist | $bin_5(dist)$ |
|---|---|---|---|---|
| 000 | lb, lbu | 00 | 0 | 00 000 |
|  |  | 01 | 24 | 11 000 |
|  |  | 10 | 16 | 10 000 |
|  |  | 11 | 8 | 01 000 |
| 001 | lh, lhu | 00 | 0 | 00 000 |
|  |  | 10 | 16 | 10 000 |
| 011 | lw | 00 | 0 | 00 000 |
| 100 | sb | 00 | 0 | 00 000 |
|  |  | 01 | 8 | 01 000 |
|  |  | 10 | 16 | 10 000 |
|  |  | 11 | 24 | 11 000 |
| 101 | sh | 00 | 0 | 00 000 |
|  |  | 10 | 16 | 10 000 |
| 111 | sw | 00 | 0 | 00 000 |

Table 6.7:  Distances of special shifts

| IR[29:28] | IR[27:26] | mnemonic | MAR[1:0] | fill |
|---|---|---|---|---|
| 00 | 00 | lb | 00 | $a_7$ |
|  |  |  | 01 | $a_{15}$ |
|  |  |  | 10 | $a_{23}$ |
|  |  |  | 11 | $a_{31}$ |
| 00 | 01 | lh | 00 | $a_{15}$ |
|  |  |  | 10 | $a_{31}$ |
| 00 | 11 | lw |  | x |
| 01 |  | lbu,lhu |  | 0 |
| 1x |  | sb,sh,sw |  | x |

Table 6.8:  Fill bit of special shifts

| IR[29,27,26] | mnemonic | mask |
|---|---|---|
| 000 | lb, lbu | $1^{24}0^8$ |
| 001 | lh, lhu | $1^{16}0^{16}$ |
| 011 | lw | $0^{32}$ |
| 1xx | sb,sh,sw | $0^{32}$ |

Table 6.9:  Mask of special shifts

address does not matter. Also, whether a load was signed or unsigned does not influence where the fill bit is applied. Thus, the mask is independent of IR[28]. It is determined according to table 6.9.

The above observations suggest a realization of the shifter environment according to the block diagram of figure 6.7. CLS(32) is a cyclic left shifter for 32-bit data. Circuit *Dist* determines the shift distance, circuit *Fill* the fill bit, and circuit *Mask* a replacement mask. Circuit *SCor* (Shift Correction) uses the mask to replace bits in the result by the fill bit. It consists of 32 1-bit multiplexers (figure 6.8). Altogether, the following formulae express cost and delay of a 32-bit shifter environment:

$$C_{SH} = C_{CLS}(32) + C_{Dist} + C_{Fill} + C_{Mask} + 32 \cdot C_{mux}$$
$$D_{SH} = \max\{D_{Dist} + D_{CLS}(32), D_{Fill}, D_{Mask}\} + D_{mux}$$

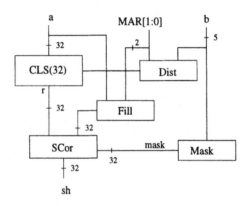

Figure 6.7: Block diagram of the shift unit SH for 32-bit input data. CLS(32) is a cyclic left shifter. Circuit Dist determines the shift distance, circuit Fill the fill bit, and circuit Mask a replacement mask. Circuit SCor uses the mask to replace bits in the result by the fill bit.

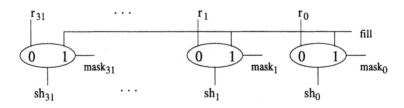

Figure 6.8: The shift-correction circuit SCor

We proceed to construct circuits Dist, Fill and Mask.

**The Distance Circuit**

The shifter environment is only used during load, store or shift operations. Thus, we can use

$$specsh = \text{IR}[31]$$

to distinguishing between explicit shifts and special shifts,

$$store = \text{IR}[29]$$

to distinguish between store and load operations, and

$$right = \text{IR}[1]$$

to distinguish between explicit right shifts and explicit left shifts.

For explicit left shifts, the shift distance comes from the $b$-input of the shifter environment. For explicit right shifts, the distance of the equivalent cyclic left shift can be obtained by inverting and incrementing the $b$-input according to lemma 3.4. The shift distance of special shifts is always a multiple of 8, thus its binary representation ends with 000. According to table 6.7, the leading two bits of the binary representation of the distance satisfy

$$bin_5(dist)[4:3] = (MAR[1], 0) \quad \text{if } MAR[0] = 0,$$

and

$$bin_5(dist)[4:3] = \begin{cases} (\overline{MAR[1]}, 1) & \text{for load} \\ (MAR[1], 1) & \text{for store} \end{cases} \quad \text{if } MAR[0] = 1.$$

Therefore, the circuit of figure 6.9 computes the shift distance at the following cost and delay:

$$\begin{aligned} C_{Dist} &= C_{Inc}(5) + 6 \cdot C_{inv} + 13 \cdot C_{mux} \\ D_{Dist} &= D_{inv} + \max\{D_{Inc}(5), D_{mux}\} + 2 \cdot D_{mux} \end{aligned}$$

**The Fill-bit Selection**

On explicit shifts, we only have a nontrivial fill bit $a_{31}$ for arithmetic right shifts, i.e., if $\text{IR}[1:0] = (1,1)$. For stores and load word (lw), the value of the fill bit does not matter, and for unsigned loads, i.e., if $\text{IR}[28] = 1$, the fill bit is 0. For the remaining 6 cases of table 6.8, we proceed in a fairly straightforward way and obtain the circuit of figure 6.10. Cost and delay obey:

$$\begin{aligned} C_{Fill} &= 7 \cdot C_{mux} + 2 \cdot C_{and} + C_{inv} \\ D_{Fill} &= \max\{3 \cdot D_{mux}, D_{inv}\} + D_{mux} + D_{and}. \end{aligned}$$

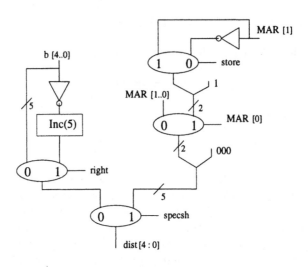

Figure 6.9: Circuit Dist which computes the shift distance

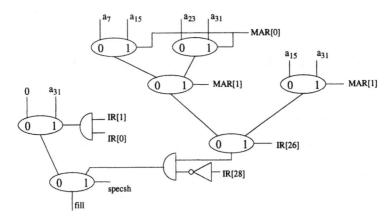

Figure 6.10: Fill-bit selection for the 32-bit shift unit

**The Mask Generation**

Mask generation for special shifts is easy. These shifts require one of the following three masks to perform the sign extension: $1^{24}0^8$, $1^{16}0^{16}$ and $0^{32}$. Table 6.9 lists the conditions under which each of these masks is used. The corresponding part of the circuit in figure 6.11 generates these masks in a straightforward way.

During an explicit left shift, the least significant $<b>$ bits have to be replaced by 0. In order to generate this mask, we use a 5-bit half decoder $hdec$ (chapter 3.3).

During an explicit right shift, the most significant $<b>$ bits have to be replaced by the fill bit. Flipping the left shift mask yields the proper mask of the right shift: this is just a matter of connecting the proper inputs and outputs. In our model, this does not cause any additional cost nor delay. Thus, the circuit of figure 6.11 generates the proper mask at the following cost and delay:

$$C_{Mask} = C_{hdec}(5) + 4 \cdot C_{mux}(32) + C_{or}$$
$$D_{Mask} = \max\{D_{hdec}(5), D_{mux}(32), D_{or}\} + 2 \cdot D_{mux}(32)$$

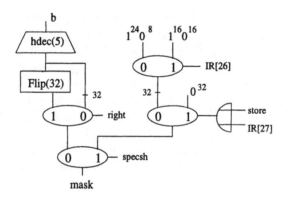

Figure 6.11: Circuit Mask. It generates the mask for a 32-bit shift.

**Connecting the Shifter Environment to the Data Paths**

Figure 6.12 depicts how to connect the shifter environment to the data paths. The $a$-input of the shifter is connected to the S1 bus, the $b$-input to the 5 least significant bits of the S2 bus and the output via a driver to the D bus. The bits IR[29:26] and IR[1:0] of the instruction register and address bits MAR[1:0] control the functionality of the shifter unit. The control unit CON only provides the enable signal $SHDdoe$ of the tristate driver. The following formulae express cost and delay of the 32-bit shifter environment:

$$C_{SHenv} = C_{SH} + C_{driv}(32)$$
$$D_{SHenv} = D_{SH} + D_{driv}(32).$$

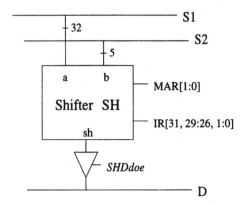

Figure 6.12: Implementation of the shifter environment

### 6.2.3  The IR-Environment

The DLX machine fetches the instruction word from memory and stores it in the instruction register IR. The instruction contains a primary and a secondary opcode, namely the fields IR[31:26] and IR[5:0]. The control logic requires these 12 bits in order to decode the instruction and to generate the proper control signals. This information has therefore to be forwarded to the control logic via the bus $Cin$. The three least significant bits of either field specify the functions which the ALU, the shifter, and the test unit have to perform. These 6 bit are passed to the corresponding units in the data paths.

The 15 bits IR[25:11] are partitioned into three fields, each of which specifies a register in the general purpose register file. These three addresses are forwarded to the register file.

For any of the following case, the instruction word provides an immediate operand which has to be extended with a fill bit to a 32-bit constant:

1. immediate arithmetic/logical operations

2. immediate test set operations

3. shift immediates

4. control operations of type I or J

The fill bit and the position of the immediate operand in the instruction word are case specific. Details are listed in table 6.10. The IR-environment of figure 6.13 uses the two control signals $Jjump$ (J-type jump) and $shift$ to distinguish between the three immediate operands. The control unit has to generate these three control signals as well as the two enable signals.

In the DLX data paths (figure 6.3), there is no direct path via a tristate driver from the busses S1 or S2 to the D bus, but that path is necessary to

| Instructions | Position | Fill Bit |
|---|---|---|
| arithmetic/logic, test set, beqz, bnez | IR[15:0] | IR[15] |
| shift | IR[10:6] | * |
| j, jal | IR[25:0] | IR[25] |

Table 6.10: Position of the immediate constant in the instruction word and value of the fill bit

execute instructions like jr or jalr. With a special constant 0 on the S2 bus, we can simulate this path by adding the data of the S1 and S2 bus and placing the result on the D bus, i.e., $D = S1 + 0$. The IR-environment therefore has an additional tristate driver which provides the special immediate operand 0. Hence, cost and delay of the IR-environment are

$$C_{IRenv} = C_{ff}(32) + 2 \cdot C_{driv}(32) + 15 \cdot C_{mux}$$
$$D_{IRenv} = D_{mux} + D_{driv}(32).$$

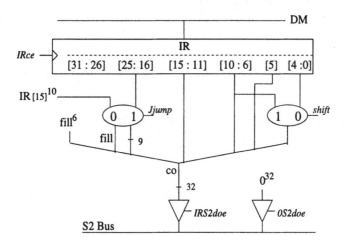

Figure 6.13: Implementation of the instruction register environment; co indicates the extended immediate operand. Control signals are printed in italics.

## 6.2.4   The GPR-Environment

The environment of the general purpose register file is shown in figure 6.14. The core of this environment is a dual-read single-write RAM. Control signal $RFw$ generated by the control unit switches between read and write mode. The RAM writes its two outputs into registers A and B and gets its input data from register C.

The observations of section 6.1.3 indicate, that the field RS1 of the instruction word (IR[25:21]) specifies the first source register and the field RS2 (IR[20:16]) the second one. But, depending on the instruction format, the address $ARw$ of the destination register is specified differently, namely

$$ARw = \begin{cases} IR[20:16] & \text{for I-type instructions} \\ IR[15:11] & \text{for R-type instructions} \\ 11111 & \text{for instructions jal and jalr} \end{cases}$$

We implement the address $ARw$ with two multiplexers selected by the control signals *Jlink* (jump and link) and *Itype*.

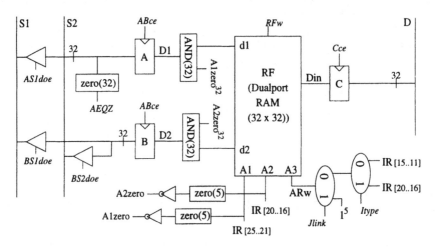

Figure 6.14: Implementation of the general purpose register environment (GPR-Environment). Control signals are printed in italics.

From the specification of the DLX instruction set (tables 6.1 - 6.3), one can conclude that register A (RS1) always serves as the first operand of an ALU or shift operation and register B (RS2) usually as the second operand. However, for store operations, register B provides the data to be written into memory; that is the first operand of the special shift operation $MDR = shift4store(B)$. The content of B must lie on the S1 bus. Consequently, the environment GPR of figure 6.14 includes three tristate drivers, one from register A to the S1 bus and two from register B to the busses S1 and S2.

According to the DLX instruction set, register R0 always provides the value 0. The general purpose register file of figure 6.14 enforces that by masking the output d1 (d2) of the dualport RAM with zero when $A1 = 0$ (respectively $A2 = 0$). For that purpose, the environment generates two signals $A1zero$ and $A2zero$ which are zero whenever the address $A1$ respectively $A2$ is zero. The RAM outputs can then be masked as follows:

$$D1 = d1 \text{ AND } A1zero = \begin{cases} d1 & \text{if } A1 \neq 0 \\ 0 & \text{otherwise} \end{cases}$$

$$D2 \; = \; d2 \text{ AND } A2zero \; = \; \begin{cases} d2 & \text{if } A2 \neq 0 \\ 0 & \text{otherwise.} \end{cases}$$

In some design systems, it is possible to change the internal structure of the RAM. That makes it much easier to guarantee $R0 = 0$. One just has to tie the data output of cell 0 to the constant 0. However, we do not use those special RAMs.

The content of register A is also tested for zero. The control unit uses this result $AEQZ$ during branch instructions. One could use the zero tester of the ALU instead of a dedicated zero tester in the GPR-environment, but this would seriously slow down the machine (section 6.5, exercise 6.1).

The environment GPR with 32 general purpose registers, each 32 bits wide, has the cost

$$\begin{aligned} C_{GPR} \; = \;\; & C_{ram2}(32, 32) + 3 \cdot C_{ff}(32) + 3 \cdot C_{driv}(32) + C_{zero}(32) \\ + \;\; & 2 \cdot C_{mux}(5) + 2 \cdot C_{zero}(5) + 2 \cdot C_{inv} + 2 \cdot C_{and}(32). \end{aligned}$$

The register file RF performs two types of accesses. It either reads out data $D1$ and $D2$, or it writes back the data from register C. The read access requires delay

$$\begin{aligned} D_{GPR}(D1, D2) \; &= \; D_{GPR}(A1, A2, RFw; D1, D2) \\ &= \; \max\{D_{ram2}(32, 32), D_{zero}(5) + D_{inv}\} + D_{and}. \end{aligned}$$

During write back, the environment selects the address $ARw$ and forwards the output of register C to the proper RAM cell. That takes time

$$D_{GPR}(; RF) \; = \; D_{GPR}(Csig, IR; RF) \; = \; 2 \cdot D_{mux}(5) + D_{ram2}(32, 32).$$

Neither time does include the time for clocking.

In this environment, further two delays are of interest, namely the delay $D_{GPR}(A, B; Sbus)$ from the registers A and B to the busses S1 and S2, and the accumulated delay of signal $AEQZ$:

$$\begin{aligned} D_{GPR}(A, B; Sbus) \; &= \; D_{driv}(32) \\ A_{GPR}(AEQZ) \; &= \; D_{zero}(32). \end{aligned}$$

## 6.2.5   The PC-Environment

The program counter PC points to the instruction to be fetched next. Since the memory is byte addressable and the instructions are 4 bytes wide, the PC has to be incremented by four during each instruction. The non-pipelined DLX machine uses the ALU to increment the PC. Instead, one could also use an additional incrementer to perform that task, but that is more expensive. The PC-environment includes two tristate drivers, one from the PC register to the S1 bus and one from the constant 4 ($0^{29}100$) to the S2 bus.

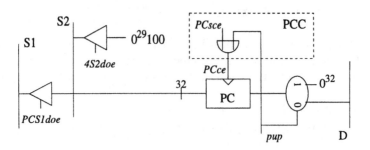

Figure 6.15: Implementation of the PC environment

After powerup, the DLX without interrupt handling must fetch the first instruction of the initial program starting at address 0. The PC environment of figure 6.15 therefore has a multiplexer selected by the external signals $pup$. However, that mechanism only works, when the control unit provides an active clock signal $PCce$ on powerup. Thus, besides the signals $PCS1doe$, $4S2doe$ the control generates signal $PCsce$ (PC standard clock) and circuit PCC (PC control) combines it with $pup$ to the clock signal $PCce$. Circuit PCC is also part of the control unit. Cost and delay of the PC environment and of circuit PCC then run at

$$\begin{aligned}
C_{PCenv} &= 2 \cdot C_{driv}(32) + C_{ff}(32) + C_{mux}(32) \\
D_{PCenv}(Sbus) &= D_{driv}(32) \\
C_{PCC} &= C_{or} \\
D_{PCC} &= D_{or}.
\end{aligned}$$

The execution of almost any DLX instruction requires an ALU or shift operation in addition to incrementing the PC. In order to avoid conflicts, both operations must be performed in different machine cycles, i.e., the execution of an instruction requires at least two cycles.

## 6.2.6   The Main Memory Environment

Figure 6.16 depicts the main memory environment and how to connect it to the data paths. Its cost runs at

$$C_{Menv} = 2 \cdot C_{ff}(32) + 2 \cdot C_{driv}(32) + C_{mux}(32).$$

Register MAR holds the address of a memory data access; only on instruction fetch, the address is directly taken from the PC. In order to make both modes possible, main memory reads the address from the tristate bus S1. That requires an additional tristate driver between register MAR and S1 bus. Register MAR can be loaded via the data bus D.

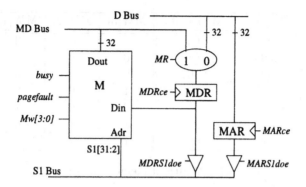

Figure 6.16: Implementation of the Memory environment

Register MDR holds the data read from memory or to be written into memory. On a read, register MDR gets its data from the memory data bus MD, otherwise it gets its data from the D bus. Therefore, the multiplexer is selected by control signal $MR$ which indicates a data read cycle. Via the tristate bus S1, register MDR provides the memory data to the data paths. Forwarding the content of registers MAR or MDR to the bus S1 takes

$$D_{Menv}(;S1) \; = \; D_{Menv}(MAR, MDR; S1) \; = \; D_{driv}(32)$$

gate delays.

Via the memory data bus MD, main memory directly provides its output data to the instruction register. This direct path is faster than the standard path through register MDR and ALU, and it speeds up the instruction fetch by one cycle.

### The Memory Control Circuit MC

Main memory sends the signals *pagefault* and *busy* to the control. The write signals $Mw[3:0]$ of the 4 banks of main memory and the signal *mis* (misalignment) are produced by the control logic in 3 steps:

1. The control automaton generates the signals $Mw$ (memory write), $MR$ (memory read) and *fetch*.

2. We observe, that read accesses including instruction fetches always access all four memory banks. For write operations, the width of the memory access and the active memory banks are determined according to table 6.11.

   Circuit MC (memory control), which is part of the control unit, generates three signals $B$, $H$ and $W$. These signals indicate the width of the write

| IR[27:26] | Width | MAR[1:0] | MW[3] | MW[2] | MW[1] | MW[0] |
|-----------|-------|----------|-------|-------|-------|-------|
| x0 | byte | 00 | 0 | 0 | 0 | 1 |
| x0 | byte | 01 | 0 | 0 | 1 | 0 |
| x0 | byte | 10 | 0 | 1 | 0 | 0 |
| x0 | byte | 11 | 1 | 0 | 0 | 0 |
| 01 | half word | 00 | 0 | 0 | 1 | 1 |
| 01 | half word | 10 | 1 | 1 | 0 | 0 |
| 11 | word | 00 | 1 | 1 | 1 | 1 |

Table 6.11: Width of memory write access

access.

$$B = \overline{IR[26]}$$
$$H = \overline{IR[27]} \wedge IR[26]$$
$$W = IR[27] \wedge IR[26]$$

Circuit MC puts the bits MAR[1:0] through a 2-bit decoder, obtains signals $B[3:0]$ and then expresses the write signals according to the following formulae:

$$Mw[0] = MW \wedge B[0]$$
$$Mw[1] = MW \wedge (W \wedge B[0] \vee H \wedge B[0] \vee B \wedge B[1])$$
$$Mw[2] = MW \wedge ((W \wedge B[0] \vee H \wedge B[2]) \vee B \wedge B[2])$$
$$Mw[3] = MW \wedge ((W \wedge B[0] \vee H \wedge B[2]) \vee B \wedge B[3])$$

3. The access is misaligned, if a halfword starts at an odd byte, if a word does not start in byte 0, or if an instruction does not start in byte 0 (PC[1:0] $\neq$ 00). MC therefore generates signal *mis* as

$$mis = (MW \vee MR) \wedge (W \wedge \overline{B[0]} \vee H \wedge (B[3] \vee B[1]))$$
$$\vee \quad fetch \wedge (PC[1] \vee PC[0])$$

Reusing the common subexpressions $W \wedge B[0]$ and $W \wedge B[0] \vee H \wedge B[2]$, the cost of circuit MC adds up to

$$C_{MC} = C_{dec}(2) + 16 \cdot C_{and} + 10 \cdot C_{or} + 3 \cdot C_{inv}.$$

Most of the computation in circuit MC is only based on the signals of the registers MAR, IR and PC; and the control signals $MW$, $MR$ and *fetch* from the bus $Csig$ are considered very late in this computation. For the output signals $Mw[3:0]$ and *mis*, we therefore specify two delays, one depending on the register inputs (Reg), and the other depending on the control inputs (Csig) of circuit MC:

$$
\begin{aligned}
A_{MC}(Reg; Mw) &= \max\{D_{dec}(2), D_{inv} + D_{and}\} + 2 \cdot D_{and} + 2 \cdot D_{or} \\
D_{MC}(Csig; Mw) &= D_{and} \\
A_{MC}(Reg; mis) &= \max\{D_{inv} + 3 \cdot D_{and} + 2 \cdot D_{or}, \\
&\qquad D_{dec}(2) + \max\{D_{or}, D_{inv}\} + 2 \cdot D_{and} + 2 \cdot D_{or}\} \\
D_{MC}(Csig; mis) &= D_{and} + 2 \cdot D_{or}.
\end{aligned}
$$

### Peculiarities of the Main Memory Design

The main memory itself is only controlled by four bank write signals $Mw[3:0]$ but not by a read signal. Thus, main memory never idles. Without an explicit request from the data paths, it performs a "dummy"-read. That does not impact the content of the main memory.

However, main memory always presents its status signals $page fault$ and $mis$ to the control. That is also not a problem, because the control knows whether these signals are relevant or not, i.e., whether the status signals correspond to an explicit memory request or not.

In order to protect the content of main memory, the control unit has to cancel a write request (inactivate the bank write signals) on a misaligned memory access. Masking the write signals by signal $mis$ is the easy way for this protection, but that slows down circuit MC and the memory environment. We therefore designed the computation of the write signals $Mw[3:0]$ in such a way, that they only become active on an aligned write access.

## 6.3   Hardwired Control

As proposed in chapter 4, we use a Moore automaton as the core of the DLX control and implement it with either of the four implementations described in that chapter, i.e., Moore vs. Mealy implementation with unary vs. binary coded states. For the control of the main memory and of the powerup mechanism in the PC environment, we extend the output circuit of the automaton by the circuits MC (section 6.2.6) and PCC (section 6.2.5).

The resulting control circuit consists of a register and two combinatorial circuits N and C, which compute the next state respectively generate the control signals. Circuit C also generates some signals to be used by circuit N. These signals are passed along bus **C2N** (C to N). When using a Moore implementation with binary coded states (the block diagram of this implementation is given in figure 6.17), bus C2N forwards the encoded state and the memory signal $mis$. Under the other three implementations of the control automaton, the bus only forwards signal $mis$.

The control unit governs the data paths and gets some signals as feedback. For these two purposes, control and data paths are connected via three busses, namely **Cin, Nin** and **Cout**. Bus Cin forwards the content of some registers of

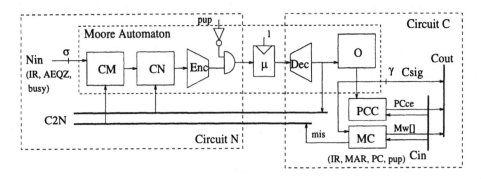

Figure 6.17: Block diagram of the hardwired DLX control, using the Moore implementation with binary coded states

the data paths and the external signals *pup* to the control circuit C. Note that in our model, external inputs are like the outputs of a register. Bus Nin forwards data from the data paths to the circuit N of the control; these data need not to come directly from registers. Bus Cout forwards the control signals to the data paths.

Since the control of our DLX design is based on a Moore automaton, bus Cin only delivers data to the circuits PCC and MC but not to the output circuit O of the automaton. All control signals of bus Cout except the clock signal $PCce$ and the write signals $MW[3:0]$ come directly from circuit O.

The signals of the three busses Cin, Nin and Cout can be derived from the circuits of the DLX data paths and its environments (section 6.2) in a straightforward way. Table 6.12 lists the signals of all three busses.

In order to estimate cost and delay of the control under any of the four implementations of the control automaton, we first have to specify a finite state diagram FSD which defines the execution scheme of the DLX instruction set. For any state in the FSD, we provide a list of active control signals. From the FSD and this list, we then derive all the parameters required to estimate cost and delay of the control automaton.

## 6.3.1 Execution Scheme

All instructions have the first two cycles in common. During **Fetch**, the first cycle, the CPU loads the instruction from the memory system. During the second cycle, the DLX increments the program counter and loads two operands from the register file. Based on the instruction opcodes, the control unit determines what sequence of cycles to start next. This cycle is therefore called *decode* or *operand fetch*.

The remaining cycles belong to the execution phase and are instruction dependent. They can be classified as computation cycles, memory accesses and register file write back. During the first execution cycle of ALU/shift opera-

| Cin | |
|-----|-----|
| MAR[1:0], PC[1:0] | position of the first data byte |
| IR[27:26] | width of the memory data access |
| pup | external signal indicating powerup |

| Nin | |
|-----|-----|
| IR[31:26], IR[5:0] | instruction opcode (IR-Env.) |
| AEQZ | test register A for zero (GPR-Env.) |
| ovf | computation overflow ($\rightarrow$ interrupt) |
| busy | memory still busy |
| pagefault | memory problems ($\rightarrow$ interrupts) |

| Cout: from circuits PCC and MC | |
|-----|-----|
| PCce | clock enable signal of the PC |
| Mw[3:0] | write signals of the 4 memory banks |

| Cout: signals $Csig$, directly from the control automaton | |
|-----|-----|
| MR, MW, fetch | type of memory access: data read / write, instruction read |
| MARce, MDRce, IRce ABce, Cce | clock enable signals |
| PCsce | standard clock enable signal of the PC |
| MARS1doe, MDRS1doe, 4S2doe, PCS1doe, IRS2doe, 0S2doe, AS1doe, BS1doe, BS2doe, SHDdoe, ALUzrDdoe, ALUzcDdoe | driver output enable signals |
| RFw | write signal of the GPR register file |
| shift | shift with immediate (IR-Env.) |
| Jjump | J-type jump (IR-Env.) |
| Jlink | jump and link (jal, jalr) (GRP-Env.) |
| Itype | I-type format (GPR-, ALU-Env.) |
| add | forces ALU to add arguments |
| test | forces AU to subtract (test set) |

Table 6.12: Signals on the three busses connecting the data paths and the control unit

tions, the hardware performs the operation and then writes the result back into the register file.

Control instructions usually end with a cycle which updates the program counter PC. Note that the PC was already incremented during the instruction decode cycle, but the branch target computation takes place several cycles later. If the branch target is $PC + \textit{offset}$ then at that point, the original PC would be required. Jumping relative to $PC + 4$ requires less hardware. The PC can always be incremented during instruction decode and the old value need not to be saved. (section 6.1.3)

The execution phase of loads and stores starts with address computation. Loads then continue with the actual load cycle, shift the memory data, and write it into the register file during the write back stage. After the address computation, stores shift the data and end by writing into memory during the store cycle. The cycles **Fetch, Load** and **Store** perform memory accesses. They must therefore be repeated till the memory ran to completion.

The finite state diagram of figure 6.18, which is almost literally from [HP90], describes the execution scheme of the DLX architecture comprising $k = 22$ states. The states of the FSD only contain the mnemonic names. The register transfer instructions (RTL instructions) of each state and the corresponding, active control signals are listed in table 6.13. We label the arc between two states $z$, $z'$ of the FSD with their disjunctive normal form (DNF) $D_i = D(z, z')$ if $D_i$ is nontrivial. In the next section, we derive these nontrivial DNFs. The number of monomials in $D(z, z')$ defines the weight of the corresponding arc $(z, z')$. For all arcs of the FSD with nontrivial DNF, table 6.14 lists the weight of the arc, all the monomials of the DNF, and the length of the monomials.

### Deriving the Nontrivial DNFs

Most of these nontrivial transitions occur in the state **Decode**, where the next state depends on the two opcodes $IR[31:26]$ and $IR[5:0]$. Here, we only show how to get the DNF of the transition from **Decode** to **TestI**. For the other transitions, the DNFs can be derived analogously.

The automaton branches from **Decode** to **TestI** on test set instructions of I-type, i.e, if the value of the primary opcode IR[31:26] lies between hx19 and hx1e (table 6.1). Thus, the disjunctive normal form $D6 := D(Decode, TestI)$ of this transition requires at least three monomials, e.g., $M6 = \{m_1, m_2, m_3\}$ with

$$
\begin{aligned}
m_1 &= \overline{IR[31]} \wedge IR[30] \wedge IR[29] \wedge IR[28] \wedge \overline{IR[27]} \\
m_2 &= \overline{IR[31]} \wedge IR[30] \wedge IR[29] \wedge \overline{IR[28]} \wedge IR[26] \\
m_1 &= \overline{IR[31]} \wedge IR[30] \wedge IR[29] \wedge IR[27] \wedge \overline{IR[26]}.
\end{aligned}
$$

However, in the current version of the DLX instruction set, no instruction is assigned to the primary opcodes hx18 and hx1f. Thus, we can use them as redundant codings of test set instructions. The I-type test set instruction then span the opcodes from hx18 to hx1f and the DNF $D6$ only comprises one

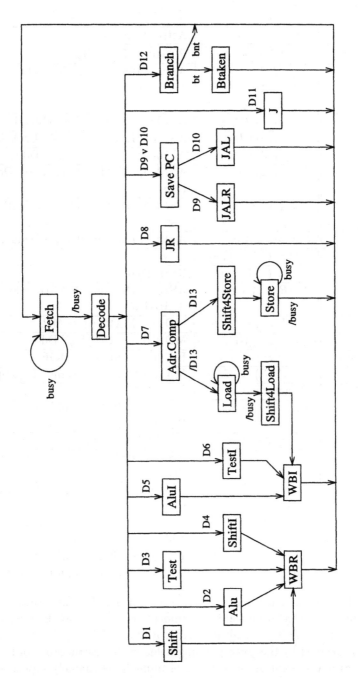

Figure 6.18: FSD of the DLX machine, comprising $k = 22$ states

| Name | RTL Instruction | Active Control Signals |
|------|-----------------|------------------------|
| Fetch | $IR = M(PC)$ | PCS1doe, fetch, IRce |
| Decode | $A = RS1,$ $B = RS2$ $PC = PC + 4$ | ABce, PCS1doe, 4S2doe, add, ALUzrDdoe, PCsce |
| Alu | $C = A$ op $B$ | AS1doe, BS2doe, Cce, ALUzrDdoe |
| Test | $C = (A$ rel $B)$ | AS1doe, BS2doe, Cce, ALUzcDdoe, test |
| Shift | $C = A$ shift $B$ | AS1doe, BS2doe, Cce, SHDdoe |
| AluI | $C = A$ op $co$ | AS1doe, IRS2doe, Cce, ALUzrDdoe, Itype |
| TestI | $C = (A$ rel $co)$ | AS1doe, IRS2doe, Cce, ALUzcDdoe, test, Itype |
| ShiftI | $C = A$ shift $cs$ | AS1doe, IRS2doe, Cce, SHDdoe, shift |
| WBR | $RD = C$ (R-type) | RFw |
| WBI | $RD = C$ (I-type) | RFw, Itype |
| Adr.Comp | $MAR = A + co$ | AS1doe, IRS2doe, MARce, add, ALUzrDdoe |
| Load | $MDR = M(MAR)$ | MDRce, MARS1doe, MR |
| Shift4Load | $C = \text{shift}(MDR)$ | MDRS1doe, Cce, ShiftDdoe |
| Shift4Store | $MDR = \text{shift}(B)$ | BS1doe, ShiftDdoe, MDRce |
| Store | $M(MAR) = MDR$ | MARS1doe, MW, (some Mw[3:0]) |
| Branch | branch taken? | |
| Btaken | $PC = PC + co$ | PCS1doe, IRS2doe, add, ALUzrDdoe, PCsce |
| J | $PC = PC + cl$ | *like Btaken,* Jjump |
| JR | $PC = A$ | AS1doe, 0S2doe, add, ALUzrDdoe, PCsce |
| Save PC | $C = PC$ | PCS1doe, 0S2doe, add, ALUzrDdoe, Cce |
| JALR | $PC = A$ $R31 = C$ | AS1doe, 0S2doe, ALUzrDdoe, add, PCsce, RFw, Jlink |
| JAL | $PC = PC + cl$ $R31 = C$ | PCS1doe, IRS2doe, ALUzrDdoe, add, PCsce, Jlink, Jjump, RFw |

Table 6.13: RTL instructions and active control signals of all states in the DLX execution scheme. $M()$ indicates a memory access, *op* an arithmetic/logical operation, and *rel* the test condition. RS1, RS2 and RD are the general purpose registers specified by the corresponding fields in the instruction word. The symbols co, cs and cl express three types of immediates, namely signed, short and long.

monomial which is even shorter than the previous ones:

$$D6 = \overline{IR[31]} \wedge IR[30] \wedge IR[29].$$

Using those redundancies means adding new instructions to the instruction set. That is problematic for two reasons:

1. In order to be binary compatible, later versions of the design must support these instructions as well.

2. The new instructions must perform admissible operations in the sense of section 2.4.5. That may require changes in the glue logic of some environments.

For the test set instructions, the three least significant bits of the opcode specify the test condition. An inspection of the comparator circuit (section 6.2.1) indicates, that opcode hx1f corresponds to an unconditional set operation, because the test condition 111, i.e., $a < b$ or $a = b$ or $a > b$, is always true. For opcode hx18 the test condition is always false, and the instruction just clears the destination register. Thus, the new instructions perform a meaningful operation.

DNFs of the transitions from **Decode** to **Test**, **Shift**, **ShiftI**, or **ALU** also make use of redundancies, adding the instructions listed in table 6.15. Table 6.14 lists all the nontrivial DNFs corresponding to the extended DLX instruction set. (Exercise 6.2 addresses the cost and run time impact of this design modification).

## 6.3.2   Cost and Delay of the Control

The analyses of chapter 4 indicate, that the parameters of table 6.16 are sufficient to estimate cost and delay of the control automaton. In a first step, we read off the values of all these parameters from the FSD and the coding of the instruction set. In a second step, we then derive the cost and the delay of the control environment.

### Parameters of the Control Automaton

Table 6.12 lists all signals of the control busses connecting the data paths and the control unit. From that, we can extract the number of input signals of circuit CM and the number of control signals generated by circuit $O$.

Circuit CM gets the bus Nin and signal $mis$ as inputs, but three of these 17 signals, namely $ovf$, $pagefault$ and $mis$, are only necessary for handling interrupts. So, circuit CM of the DLX without interrupt handling has $\sigma = 14$ relevant input signals.

The control unit provides 31 control signals on the bus Cout, but four of these signals are generated by circuit $MC$. So, the control automaton only generates $\gamma = 28$ of the control signals.

The FSD of the DLX architecture (figure 6.18) comprises $k = 22$ states which can be coded in binary with $\zeta = 5$ bits. For each state of the FSD, table 6.13 lists the active control signals. Counting the state of the FSD in which a control

| Nontrivial DNF | Target State | Monomial $m \in M$ | | Length $l(m)$ | Weight of arc |
| | | IR[31 : 26] | IR[5 : 0] | | |
|---|---|---|---|---|---|
| D1 (*) | Shift | 000000 | 0001** | 10 | 1 |
| D2 (*) | Alu | 000000 | 100*** | 9 | 1 |
| D3 (*) | Test | 000000 | 101*** | 9 | 1 |
| D4 (*) | ShiftI | 000000 | 0000** | 10 | 1 |
| D5 | AluI | 001*** | ****** | 3 | 1 |
| D6 (*) | TestI | 011*** | ****** | 3 | 1 |
| D7 | Adr.Comp | 100*0* | ****** | 4 | 3 |
| | | 10*0*1 | ****** | 4 | |
| | | 10*00* | ****** | 4 | |
| D8 | JR | 010110 | ****** | 6 | 1 |
| D9 | JALR | 010111 | ****** | 6 | 1 |
| D10 | JAL | 000011 | ****** | 6 | 1 |
| D9 ∨ D10 | Save PC | like D9 and D10 | | | 2 |
| D11 | J | 000010 | ****** | 6 | 1 |
| D12 | Branch | 00010* | ****** | 5 | 1 |
| D13 | Shift4Store | **1*** | ****** | 1 | 1 |
| /D13 | Load | **0*** | ****** | 1 | 1 |
| bt | Btaken | AEQZ ·/IR[26] | | 2 | 2 |
| | | /AEQZ ·IR[26] | | 2 | |
| bnt | Fetch | /AEQZ ·/IR[26] | | 2 | 2 |
| | | AEQZ ·IR[26] | | 2 | |
| busy | | busy | | 1 | 1 |
| /busy | | /busy | | 1 | 1 |
| Accumulated length of $m \in M$: $\sum_{m \in M} l(m)$ | | | | 97 | |

Table 6.14: Nontrivial disjunctive normal forms (DNF) of the DLX finite state diagram and the corresponding monomials. DNFs marked with a star profit from the extended instruction set.

| IR[31 : 26] | IR[5 : 0] | Mnemonic | Effect |
|---|---|---|---|
| R-type Instructions | | | |
| hx00 | hx01 | slai | RD = RS1 $\ll_{\text{arith}}$ SA |
| hx00 | hx05 | sla | RD = RS1 $\ll_{\text{arith}}$ RS2[4 : 0] |
| hx00 | hx27 | lh | RD = RS2[15:0] $0^{16}$ |
| hx00 | hx28 | clr | RD = ( false ? 1 : 0); |
| hx00 | hx2f | set | RD = ( true ? 1 : 0); |
| I-type Instructions | | | |
| hx18 | hx** | clri | RD = ( false ? 1 : 0); |
| hx1f | hx** | seti | RD = ( true ? 1 : 0); |

Table 6.15: Extension of the DLX Instruction Set. All instructions also execute *PC += 4.*

| Symbol | Meaning | Value |
|---|---|---|
| $\sigma$ | # inputs of CM | 14 |
| $\gamma$ | # output signals of O | 28 |
| $k$ | # states of FSD | 22 |
| $\zeta$ | $\zeta = \lceil k \rceil$ | 5 |
| $\nu_{max}$ | maximal frequency of a control signal in FSD | 10 |
| $\nu_{sum}$ | accumulated frequency of all control signals | 93 |
| $\#M$ | # monomials $m \in M$ (nontrivial) | 22 |
| $l_{max}$ | length of longest monomial $m \in M$ | 10 |
| $l_{sum}$ | accumulated length of all monomials $m \in M$ | 97 |
| $\text{fanin}_{max}$ | maximal fanin of nodes ($\neq$ fetch) in FSD | 4 |
| $\text{fanin}_{sum}$ | accumulated fanin | 32 |

Table 6.16: Parameters relevant for cost and delay of the control automaton

| MR | 1 | MW | 1 | MARce | 1 | MDRce | 2 |
|---|---|---|---|---|---|---|---|
| MARS1doe | 2 | MDRS1doe | 1 | IRce | 1 | PCsce | 6 |
| ABce | 1 | Cce | 8 | IRS2doe | 7 | 0S2doe | 3 |
| PCS1doe | 6 | 4S2doe | 1 | ALUzrDdoe | 10 | ALUzcDdoe | 2 |
| AS1doe | 9 | BS1doe | 1 | BS2doe | 3 | ShiftDdoe | 4 |
| RFw | 4 | shift | 1 | add | 8 | fetch | 1 |
| Jlink | 2 | Jjump | 2 | Itype | 3 | test | 2 |

Table 6.17: Frequency of the active control signals in the FSD

| Fanin | States of FSD |
|-------|---------------|
| 11 | fetch $(= z_0)$ |
| 4 | WBR |
| 3 | WBI, Adr.Comp |
| 2 | SavePC, Load, Store, Btaken |
| 1 | remaining 14 states |
| 32 | $\text{fanin}_{sum}$, accumulated fanin of states $\neq$ fetch |

Table 6.18: Fanin of all states in the FSD of the DLX

signal $O_i$ is active yields the frequencies $\nu_i$ listed in table 6.17. Signal $ALUzrDdoe$ has the highest frequency of

$$\nu_{max} := \max\{\nu_i \,|\, 1 \leq i \leq \gamma\} = 10.$$

Summing up the frequency $\nu_i$ of all the 28 control signals yields

$$\nu_{sum} := \sum_{1 \leq i \leq 28} \nu_i = 93.$$

Table 6.14 lists all the nontrivial monomials $M$ of the FSD. For the DLX design, set $M$ contains $\#M = 22$ nontrivial monomials. The longest monomials D1 and D4 comprise

$$l_{max} := l(D1) = l(D4) = 10$$

literals. Summing up the length of all the nontrivial monomials yields

$$l_{sum} := \sum_{m \in M} l(m) = 97.$$

Arcs in the FSD with trivial DNF have weight 1. For all the other arcs, table 6.14 lists their weight. In the next step, we compute the fanin of any node $z$ in the FSD as the accumulated weight of all arcs ending in node $z$. Table 6.18 lists the fanin of all states. State WBR has the highest fanin of all states different from the initial state fetch $(z_0)$, namely

$$\text{fanin}_{max} := \max\{\text{fanin}(z) \,|\, z \neq z_0\} = 4.$$

The fanin of all states different from fetch add up to

$$\text{fanin}_{sum} := \sum_{z \neq z_0} \text{fanin}(z) = 32.$$

**Cost and Delay of the Control Environment**

The hardwired control of the DLX consists of the control automaton and the circuits PCC and MC. Let

$$C_{auto} = \begin{cases} C_{Moore} & \text{, Moore impl., binary coded states} \\ C_{uMoore} & \text{, Moore impl., unary coded states} \\ C_{Mealy} & \text{, Mealy impl., binary coded states} \\ C_{uMealy} & \text{, Mealy impl., unary coded states} \end{cases}$$

be the cost of the control automaton. The cost of the control environment then runs at

$$C_{CON} = C_{PCC} + C_{MC} + C_{auto}.$$

Circuit C of the control environment computes all control signals. Besides the state of the automaton, it only gets inputs via bus Cin which forwards the content of some registers in the data paths. Circuit C consists of circuit PCC, which generates clock signal $PCce$, of circuit MC, which generates signal $mis$ and the write signals $Mw[3:0]$ of main memory, and of the output circuit of the control automaton, which generates the remaining control signals $Csig$. It takes the automaton

$$A_{CON}(Csig) = \begin{cases} A_{Moore}(Csig) & \text{, Moore impl., binary coding} \\ A_{uMoore}(Csig) & \text{, Moore impl., unary coding} \\ A_{Mealy}(Csig) & \text{, Mealy impl., binary coding} \\ A_{uMealy}(Csig) & \text{, Mealy impl., unary coding} \end{cases}$$

gate delays to generate its control signals $Csig$. These signals are also inputs of PCC and MC. Thus, the delays of the control signals $PCce$, $Mw[3:0]$ and $mis$ run at

$$\begin{aligned} A_{CON}(PCce) &= A_{CON}(Csig) + D_{PCC} \\ A_{CON}(Mw) &= \max\{A_{CON}(Csig) + D_{MC}(Csig; Mw), A_{MC}(Reg; Mw)\} \\ A_{CON}(mis) &= \max\{A_{CON}(Csig) + D_{MC}(Csig; mis), A_{MC}(Reg; mis)\}. \end{aligned}$$

# 6.4  Hardware Cost and Cycle Time

In the previous sections, we developed formulae which estimate the cost and the delay of all units of the DLX hardware. We now evaluate the corresponding C-programs (see appendix B.6) and determine the cost and the cycle time of the DLX hardware under the given technology. Based on those figures, we perform some design optimizations.

| Environment | ALU | SH | IR | GPR | PC | M | DP |
|:---:|:---:|:---:|:---:|:---:|:---:|:---:|:---:|
| Motorola | 1858 | 1338 | 621 | 5406 | 672 | 928 | 10823 |
| Venus | 2182 | 1378 | 813 | 21627 | 864 | 1248 | 28112 |

Table 6.19: Cost of the DLX data paths and of all its environments

| Implementation | | Moore | | Mealy | | Register |
|:---:|:---:|:---:|:---:|:---:|:---:|:---:|
| Coding of the states | | unary | binary | unary | binary | $C_{ff}(\gamma)$ |
| Motorola | $C_{CON}$ | 792 | 787 | 1014 | 1083 | 224 |
| technology | $C_{DLX}$ | 11615 | 11610 | 11837 | 11906 | |
| Venus | $C_{CON}$ | 880 | 807 | 1214 | 1215 | 336 |
| technology | $C_{DLX}$ | 28992 | 28919 | 29326 | 29327 | |

Table 6.20: Cost of the hardwired DLX control and of the DLX hardware

## 6.4.1 Hardware Cost

**Data Paths** The data paths (DP) of the non-pipelined DLX without interrupt handling consists of six environments, namely: arithmetic logic unit (ALU), shifter (SH), instruction register (IR), general purpose registers (GPR), program counter (PC), and memory (M). The cost of the 32-bit data paths is

$$C_{DP} = C_{ALUenv} + C_{SHenv} + C_{IRenv} + C_{GPR} + C_{PCenv} + C_{Menv}.$$

Table 6.19 lists the cost of the data paths and all its environments under Motorola technology [NB93] and Venus technology [HNS86].

Under Motorola technology, the environment of the general purpose register roughly accounts for 50% of the cost of the 32-bit DLX data paths. Of course, this fraction depends on the number of general purpose registers used. The older Venus technology only provides expensive RAMs at 12 gate equivalents per cell. Under this technology constraint, the general purpose register environment even accounts for 76% of the cost of the DLX data paths.

**Control Unit** As core of the hardwired DLX control, we use a Moore automaton. For this automaton, we provide four implementations, namely a Moore / Mealy implementation with binary/unary coded states. Table 6.20 lists the cost $C_{CON}$ of the control and the cost $C_{DLX}$ of the DLX hardware for all four implementations under both technologies.

Compared to the Moore implementation, the Mealy implementation of the control automaton requires additional $\gamma = 28$ flipflops, and when coding the states in binary, even an $(k-1)$-bit zero tester and some AND gates. Under Motorola technology, switching from a Moore implementation to a Mealy implementation therefore increases the cost of the control automaton with states coded in unary (binary) by 28% (38%). For the Venus technology, the cost increase is even about 10% higher, due to expensive registers.

In accordance with the general results of section 4.4, for the FSD with $k = 22$ states, coding the states in binary reduces the cost of the automaton about 8% on a Moore implementation but increases the cost up to 7% on a Mealy implementation.

The figures of table 6.20 also support the following rule of thumb of Engineering:

**Maxim 6.1 (Cost of Hardwired Control)** *The hardwired control of a non-pipelined DLX without interrupt handling accounts for less than 10% of the total cost of the DLX machine with 32-bit wide data paths and 32 general purpose registers.*

For technologies with expensive storage components like VENUS – i.e., a flipflop and a RAM cell require 12 gate equivalents – even a stronger version of the maxim holds. The cost of the control then even stays under 5% of the total hardware cost of the 32-bit DLX.

Changes in the implementation of the control automaton, like switching from unary to binary coded states, or replacing a Moore implementation by a Mealy implementation, have virtually no impact on the total cost of the DLX (less than 2.6%).

## 6.4.2   Cycle Time

Computing the cycle time of the DLX architecture is a three step process. In the first step, we determine the delay of the control signals on the bus Cout. These signals govern the data paths of the DLX architecture. In the second step, we compute the cycle time of the data paths and the accumulated delay of the signals Nin which the data paths forward to the next state circuit of the control environment. In the third step, we finally determine the cycle time of the control environment and the cycle time of the whole DLX design.

### Delay of the Control Signals

According to section 6.3.2, at the beginning of a new cycle, the control automaton requires $A_{CON}(Csig)$ gate delays to generate the current control signals $Csig$. Past that time, the control signals do not toggle for the rest of the cycle; we say the control signals are *valid*. The write signals $Mw[3:0]$ and the clock signal $PCce$ are valid $A_{CON}(Mw)$ respectively $A_{CON}(PCce)$ delays after the start of the cycle.

Table 6.21 lists these three delays for all four implementations of the hardwired DLX control under both technologies. Using a Mealy approach, the *standard* control signals, i.e., the control signals $Csig$, are directly taken from a register in the control unit. Therefore, the standard control signals are valid immediately after the new cycle started and switching between binary and unary coding of the states has no impact on the run time.

| Technology | Approach | Moore | | Mealy | |
|---|---|---|---|---|---|
| | Coding | unary | binary | unary | binary |
| Motorola | $A_{CON}(Csig)$ | 8 | 15 | 0 | 0 |
| | $A_{CON}(Mw)$ | 14 | 21 | 11 | 11 |
| | $A_{CON}(PCce)$ | 10 | 17 | 2 | 2 |
| Venus | $A_{CON}(Csig)$ | 4 | 8 | 0 | 0 |
| | $A_{CON}(Mw)$ | 7 | 11 | 6 | 6 |
| | $A_{CON}(PCce)$ | 5 | 9 | 1 | 1 |

Table 6.21: Delay of the control signals

Using a Moore implementation, the standard control signals first have to pass the output circuit OD. With unary coded states, that delays the control signals by 8 respectively 4 gate delays. When binary coding the states, circuit OD also contains a decoder. That roughly doubles the delay of the standard control signals.

The write signals of main memory have to pass the memory control circuit $MC$ which has some standard control signals as inputs. Since the computation of circuit MC and of the control automaton can partially be overlapped, the difference $A_{CON}(Mw) - A_{CON}(Csig)$ is not a constant. Since the computation of circuit PCC and of the control automaton can not be overlapped, for signal $PCce$ their delays do add.

### Cycle Time of the Data Paths

To compute the cycle time of the data paths, we have to consider all paths starting in a register, in a register like constant or with an external signal and ending in a register or the register file (RAM). In order to spot all cycles, we use a simplified view of the data paths (figure 6.19). For the computation of the cycle time of a cycle, we then take a closer view at the underlying circuits.

**Simplified View of the Data Paths**  There are three types of storage components in the data paths, namely: the registers, the register file RF, and the memory M. The latter two can be modeled as a combination of a read circuit, of a write circuit and of a register. By register $Reg$, we denote the registers PC, IR, A, B, C, MAR and MDR. Register Reg is connected to the busses S1 and S2 via a circuit RegS. Thus, circuit RegS combines circuits from the environments of the general purpose register file, of the program counter, of the instruction register and of the memory. The same holds for circuit $Regw$ which connects register Reg to the data busses D and MD. Circuit $RegCE$ collects the clock enable signals of the registers $Reg$ from the control automaton and from circuit PCC. All circuits in the data paths of the DLX architecture are governed exclusively by the control circuits $Csig$ or by external signals ($pup$), only the memory receives additional control signals from circuit MC.

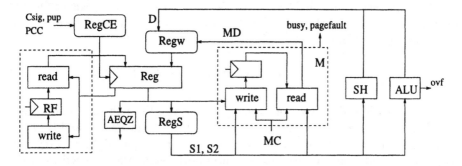

Figure 6.19: Simplified view of the DLX data paths

From figure 6.19, one can easily read off the cycles through the data paths:

1. through RegCE into register Reg,

2. from register Reg into the register file RF,

3. from the register file RF back into register Reg,

4. from register Reg through the ALU into Reg,

5. from register Reg through the shifter SH into Reg,

6. from register Reg through the memory M into Reg, and

7. from register Reg into the memory M.

**Updating Registers Reg**  The control automaton computes almost all the clock enable signals of the registers in the DLX data paths, but signal $PCce$ is generated by control circuit PCC. That requires $A_{CON}(PCce)$ gate delays. Thus, the delay of the clock signals requires a cycle time of at least $T_{RegCE}$ gate delays, with

$$T_{RegCE} = \max\{A_{CON}(Csig), A_{CON}(PCce)\} + \Delta.$$

**Register File Accesses**  The next two paths involve the dual-port RAM of the register file, one for write and one for read. The close-up of the register file (figure 6.14 on page 95) indicates that during write back, the registers C and IR provide the data and the address. The address has to pass a circuit governed by control signals from bus $Csig$, before it reaches the register file. This circuit and the register file have together a delay of $D_{GPR}(; RF)$. Thus, the write back cycle of the register file takes

$$T_{GRF}(; RF) = A_{CON}(Csig) + D_{GPR}(; RF) + \delta.$$

This already includes the time overhead for clocking. During read, the address comes directly from the instruction word and the data D1 and D2 are written into the registers A and B. The control signals $Csig$ switch the register file into read mode and provide the clock signals of the registers A and B. The read cycle therefore requires time:

$$T_{GPR}(RF;) = A_{CON}(Csig) + D_{GPR}(D1, D2) + \Delta.$$

**Circuits RegS and Regw** All the other paths go through the circuits RegS and Regw, i.e., they start with placing data on the S1 and S2 bus, and they end with saving the result from the D bus or the memory output MD into a register. The cycle of the memory write access starts in the same way, but writes in the memory instead.

All environments with drivers on the S1 or S2 bus contribute to circuit RegS, and impact its accumulated delay $A_{RegS}$. The relevant environments include the program counter, the instruction register, the register file and main memory. Output signals from the control automaton govern all these drivers and the circuits which deliver the data to the drivers. That consequently results in the following delay

$$A_{RegS} = A_{CON}(Csig) + \max\{D_{IRenv}, D_{PCenv}(Sbus),$$
$$D_{GPR}(A, B; Sbus), D_{Menv}(; S1)\}.$$

Circuit Regw forwards the results from the data busses D and MD to the registers Reg of the data paths. That involves the same environments as circuit RegS. The data inputs of almost all registers are directly connected to D or MD. The only exceptions are the registers PC and MDR of the PC and memory environment (figures 6.15, 6.16), where the data have to pass a multiplexer. These multiplexers are controlled by external signal $pup$ respectively by a standard control signal. Since both signals are valid before the data arrive, the write process has delay

$$D_{Regw} = D_{mux}(32).$$

**Shift and Compute Cycles** During shift cycles, the data from the S1 and S2 bus are passed through the shifter environment on to the D bus and are saved in registers. The instruction word and signals $Csig$ control the shifter environment. These signals are valid before the data arrive at the shifter. Shift cycles therefore take at most time

$$T_{SH} = A_{RegS} + D_{SH} + D_{Regw} + \Delta.$$

The ALU environment receives data inputs via the busses S1 and S2 and control inputs via bus $Csig$. It then generates the result $zr$ and the condition $zc$, and forwards one of them to the bus D. In section 6.2.1, we specified for either output two delays, one from the busses S1 and S2 to the output and one from the control inputs $Csig$ to the output. Thus, the cycle time of the two ALU cycles can be expressed as

$$
\begin{aligned}
T_{ALU}(ZR) &= max\{A_{CON}(Csig) + D_{ALUenv}(Csig; ZR),\\
&\quad A_{RegS} + D_{ALUenv}(S1, S2; ZR)\} + D_{Regw} + \Delta
\end{aligned}
$$

$$
\begin{aligned}
T_{ALU}(ZC) &= max\{A_{CON}(Csig) + D_{ALUenv}(Csig; ZC),\\
&\quad A_{RegS} + D_{ALUenv}(S1, S2; ZC)\} + D_{Regw} + \Delta
\end{aligned}
$$

The ALU also generates an overflow signal within

$$
\begin{aligned}
A_{ALU}(ovf) &= max\{A_{CON}(Csig) + D_{ALUenv}(Csig; ovf),\\
&\quad A_{RegS} + D_{ALUenv}(S1, S2; ovf)\}
\end{aligned}
$$

gate delays after the start of a new cycle. Note, that on a DLX without interrupt handling this signal deadends.

**Main Memory Accesses**   The remaining two cycles of the DLX data paths are main memory accesses. According to the figures 6.19 and 6.16, the S1 bus provides the memory address, register MDR provides the data to be written into memory, and control circuit MC provides the write signals $Mw[3:0]$.

During a read access (load or instruction fetch), data arrive $d_{mem}$ delays after all memory inputs are valid on the bus MD and can be forwarded to a register via circuit Regw. Thus, the whole memory read access has the cycle time

$$
T_M(M;) = max\{A_{CON}(Mw), A_{RegS}\} + d_{mem} + D_{Regw} + \Delta.
$$

A write access basically starts like a read access, and register MDR provides the input data directly to the main memory. The cycle time of a write access therefore is

$$
T_M(;M) = max\{A_{CON}(Mw), A_{RegS}\} + d_{mem} + \delta.
$$

Thus, writes are faster than reads, and the cycle time of memory accesses equals

$$
T_M = T_M(M;).
$$

The memory environment also generates two status signals *pagefault* and *busy*, but without interrupt handling only signal *busy* is forwarded to the control unit. With a memory status time of $d_{mstat}$, the two status signals become valid $A_{Menv}(mstat)$ delays after the cycle started, with

$$
A_{Menv}(mstat) = max\{A_{CON}(Mw), A_{RegS}\} + d_{mstat}.
$$

For the later analysis, we assume a memory status time of $d_{mstat} = 5$ gate delays.

| Cycle time | | GPR | | SH | ALU | | DP |
| | Implementation | (;RF) | (RF;) | | (ZR) | (ZC) | |
|---|---|---|---|---|---|---|---|
| Motorola | Moore, unary | 33 | 35 | 46 | 75 | 88 | 88 |
| | Moore, binary | 40 | 42 | 53 | 82 | 95 | 95 |
| | Mealy | 25 | 27 | 38 | 67 | 80 | 80 |
| Venus | Moore, unary | 21 | 21 | 37 | 47 | 53 | 53 |
| | Moore, binary | 25 | 26 | 41 | 51 | 57 | 57 |
| | Mealy | 17 | 18 | 33 | 43 | 49 | 49 |

Table 6.22: Cycle time of the DLX data paths and delay of the major cycles through the data paths

**Cycle Time of DP**   The cycle time $T_{DP}$ is the maximal time of all cycles in the DLX data paths except those through the main memory:

$$T_{DP} = \max\{T_{RegCE}, T_{GPR}(; RF), T_{GPR}(RF;),$$
$$T_{SH}, T_{ALU}(ZR), T_{ALU}(ZC)\}.$$

Table 6.22 lists the cycle time of the DLX data paths and of all the relevant cycles within the data paths. All cycles except the read cycle of the register file first consider the standard control signals. Thus, the accumulated delay $A_{CON}(Csig)$ adds to the duration of almost all the cycles through the data paths. Using a Mealy implementation of the control automaton therefore speeds up the cycle time of the data paths in the same way than the computation of the control signals.

The time critical path goes through the adder and the comparator. Under Motorola (Venus) technology, the comparator accounts for 18% (13%) of the delay of this critical path. This cycle is over 39% slower than any other cycle in the data paths which avoids the adder. In chapter 7, we therefore analyze how to speed up this path.

**Cycle Time of the Control**

As the block diagram of the control environment (figure 6.17) indicates, only the control automaton updates the state of the control unit. Thus, the cycle time of the control unit equals the cycle time of the control automaton. According to the formulae of section 6.3.2, this cycle time can be expressed as

$$T_{CON} = \begin{cases} T_{Moore} & \text{, Moore impl., binary coded states} \\ T_{uMoore} & \text{, Moore impl., unary coded states} \\ T_{Mealy} & \text{, Mealy impl., binary coded states} \\ T_{uMealy} & \text{, Mealy impl., unary coded states} \end{cases}$$

assuming that the inputs $In$ of control circuit N, i.e., the signals on bus Nin and signal $mis$, have an accumulated delay of $A(In)$. In the current DLX design

| | Implementation | Acc. Delay | | Cycle Time | | |
|---|---|---|---|---|---|---|
| | | AEQZ | busy | $T_{CON}$ | $T_{DP}$ | $T_{Menv}$ |
| M | Moore, unary | 9 | 14 | 45 | 88 | 21 |
| | Moore, binary | 9 | 21 | 53 | 95 | 28 |
| | Mealy, unary | 9 | 11 | 44 | 80 | 18 |
| | Mealy, binary | 9 | 11 | 44 | 80 | 18 |
| V | Moore, unary | 5 | 7 | 26 | 53 | 14 |
| | Moore, binary | 5 | 11 | 30 | 57 | 18 |
| | Mealy, unary | 5 | 6 | 26 | 49 | 13 |
| | Mealy, binary | 5 | 6 | 26 | 49 | 13 |
| Fixed Displacement | | | | $+d_{mstat}$ | | $+d_{mem}$ |

Table 6.23: Cycle time of the control, the data paths and the memory environment. M indicates Motorola technology and V Venus technology.

without interrupt handling, signal *mis* (C2N bus) of the memory control circuit $MC$ does not impact the behavior of the control. Therefore, the state $\mu$ and the signals from bus Nin are the only inputs of circuit N. The bus Nin forwards data from the data paths. Most of them are supplied directly by registers or a register like external signal (*pup*), but not so the signal $AEQZ$ of the GPR-environment and the signal *busy* of the memory environment. Thus, the input data $In$ of circuit N get valid

$$A(In) = \max\{A_{GPR}(AEQZ), A_{Menv}(mstat)\}$$

delays after the cycle started.

**Cycle Time of the DLX Machine**

The cycle time $T_{DLX}$ of the non-pipelined DLX design is the maximum of the cycle time of the data paths and of the control environment:

$$T_{DLX} = \max\{T_{DP}, T_{CON}\}.$$

The cycle time $T_M$ of the memory environment only impacts indirectly the cycle time of the DLX machine, for the following reason. If the memory cycle time is less than the cycle time $T_{DLX}$, memory accesses can be performed in one machine cycle. In the other case, $T_M \geq T_{DLX}$, the cycle time of the machine must be increased to $T_M$ or memory accesses are performed in $\lceil T_M / T_{DLX} \rceil$ cycles. Our design goes along the second line.

Table 6.23 lists the cycle time of the data paths, of the control environment and of the memory environment. These figures indicate, that in the current DLX design, memory accesses can be performed in one cycle if the memory access time is less than 68% of the cycle time of the DLX data paths.

Among others, the table also lists the cycle time of the control and the delay of its two input signals with nontrivial delay, $AEQZ$ and $busy$. Under the current conditions, signal $busy$ has a longer delay than signal $AEQZ$, and the memory status time $d_{mstat}$ therefore directly adds to the cycle time $T_{CON}$ of the control environment. On the current DLX design, the control does not dominate the cycle time of the machine if the memory status time stays under 45% of the cycle time of the DLX data paths. However, in exercise 6.4 we discuss an optimization of the DLX control automaton, which allows us to tolerate a memory status time of 65% to 84% of $T_{DP}$.

On a Mealy implementation of the hardwired control, the standard control signals are directly taken from a register of the control automaton. Therefore, a stronger version of the above rule of thumb holds:

**Maxim 6.2 (Cycle Time of Hardwired Control)** *If the hardwired control of the non-pipelined DLX without interrupt handling is realized as Mealy implementation of a Moore automaton, and if the memory status time $d_{mstat}$ is small, the control does not lie on the time critical path of the design, even when the automaton is implemented without any optimization.*

Engineers design the control of their RISC designs on a high level (FSD) and use tools to generate the implementation of the control unit. That is similar to what we do for our cost and delay estimate. Since the so generated control logic usually does not lie on the time critical path, that is the right way to go to avoid a lot of error prone work.

### 6.4.3 Design Evaluation

Let us assume that the memory status time stays under 46% of $D_{DP}$; that is a reasonable assumption for on-chip memory. Then, under Motorola (Venus) technology, the DLX design can run at a cycle time $t_C \geq 80$ (49) delays. According to table 6.23, a Moore implementation of the control automaton slows the DLX down by 4-15 delays. That is roughly 10 to 20% of the cycle time. On the other hand, a Moore implementation of the control automaton reduces the cost of the DLX by about 2.6%. Consequently, the DLX with Mealy implementation has a better price/performance ratio (7-18% better) and this is independent of the benchmark application.

The Mealy control automaton which codes the states in unary results in the most cost efficient DLX design. For the rest of this section, we denote this implementation of the control automaton as the unary Mealy automaton. Diagram 6.20 depicts the quality ratio of the DLX designs with Moore control automaton relative to the DLX design with unary Mealy control automaton as a function of the quality parameter $q$. These graphs indicate, that the Moore variants of the DLX have a better quality than the Mealy variant only if the cost has a much stronger impact on the quality $Q_q$ than the run time; the quality parameter $q$ must be greater than 0.83. However, for a quality metric which we consider to be realistic (see chapter 2.3), the parameter $q$ lies in the range $[0.2, 0.5]$. Consequently, the following rule of thumb holds:

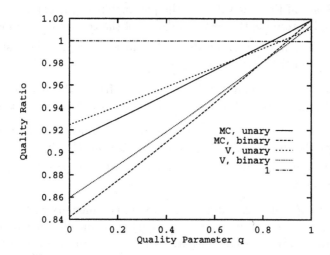

Figure 6.20: Quality ratio of the DLX designs with Moore based control automaton relative to the DLX design with unary Mealy control automaton. $MC$ indicates Motorola technology and V Venus technology

**Maxim 6.3** *Under a realistic quality metric $Q_q$ with $q \in [0.2, 0.5]$, the design variant of the non-pipelined DLX without interrupt handling which uses a Mealy implementation of the control automaton has a higher quality than the design variants which use a Moore implementation of the control automaton.*

## 6.5   Exercises

6.1 The GPR environment of section 6.2.4 provides an 32-bit zero tester in order to test source register RS1. However, the comparator circuit of the ALU can also perform this test. Thus, dropping the additional zero tester only changes the data paths used while executing state **Branch** of the FSD (figure 6.18). Analyze, how this modification impacts the design, the cost, and the cycle time of the DLX architecture.

   1. Update the list of active control signals.

   2. Derive the values of the parameters relevant for the cost and delay of the control automaton.

   3. Determine cost and cycle time of the modified DLX design.

6.2 In order to simplify the DNFs of 5 transitions in the control automaton of the DLX machine, we have added 7 new instructions (table 6.15).

   1. Derive the DNFs of these 5 transitions under the original DLX instruction set.

2. Read off the new values of the parameters of the control automaton, and determine the cost and delays of the control environment.

3. By how much does this design modification impact the cost and cycle time of the DLX machine?

6.3 Analyze how switching the fields RS2 and RD in the R-type instruction format impacts the DLX design (see section 6.1.3). Proceed as follows:

1. Modify the glue logic of the GPR environment and derive formulae for the the delays $D_{GPR}(; RF)$ and $D_{GPR}(D1, D2)$.

2. Does this modification impacts the cost and cycle time of the DLX machine?

6.4 The data of table 6.23 indicate, that for a main memory with a large status time $d_{mstat}$ the computation of of signal $busy$ and the control automaton lie both on the time critical path. However, on a memory access, signal $busy$ forces the automaton to return to the current state (FSD, figure 6.18).

The modified control automaton of figure 6.21 speeds up the critical paths. Its output circuit must generate signal $memory$ which indicates a memory access. The select signal $sel$ of the new multiplexer is then generated as

$$sel = (memory \land \overline{pup}) \land busy.$$

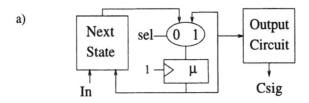

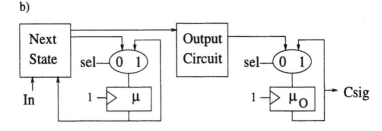

Figure 6.21: Modification of the DLX control automaton using a Moore implementation (a) respectively a Mealy implementation (b)

1. Derive cost and delay formulae for the Moore / Mealy implementation of the modified DLX control automaton and of the control environment.

2. How large a memory status time can the modified DLX design tolerate?

# Chapter 7

# Trade-Off Analyses

In this chapter, we study ways to speed up the DLX machine and analyze the impact of the changes on the quality of the design. Those design evaluations were the main incentive for us to introduce a cost and delay function into the architecture model. Modifications which reduce the cost and the run time are most desired. However, modifications usually improve the run time but increase the cost or vice versa. In this context, the extended architecture model enables the quantification of so called *cost / run time trade-offs*.

The run time analysis of section 6.4.2 indicates, that the delay of the path through the arithmetic unit AU and the comparator dominates the cycle time of the DLX, and that any path which avoids the arithmetic unit is faster by far. We therefore choose the following two methods to speed up the delay of the critical path in the DLX design, namely:

- different implementations of the adder

- different implementations of the condition test

In section 7.3 we then evaluate the design variants of the non-pipelined DLX design.

## 7.1 Modification of the Arithmetic Unit

The core of the arithmetic unit AU of the DLX is the adder. In the original design, we used a *carry look-ahead adder* (CLA). Now, we replace it either by the *ripple carry adder* (section 3.7.2) or by the *conditional sum adder* (section 3.7.4).

The arithmetic unit AU is still composed in the same manner as in the original DLX design (section 3.7, figure 7.1), but now it is based on various adders. Cost and delay of the modified arithmetic unit therefore run at

$$
C_{AU}(n) = C_{xor}(n) + \begin{cases} C_{CLA}(n) & \text{, carry look-ahead} \\ C_{RCA}(n) & \text{, ripple carry} \\ C_{CSA}(n) & \text{, conditional sum} \end{cases}
$$

| Condition | | false | $a > b$ | $a = b$ | $a \geq b$ | $a < b$ | $a \neq b$ | $a \leq b$ | true |
|---|---|---|---|---|---|---|---|---|---|
| $f_2$ | $<$ | 0 | 0 | 0 | 0 | 1 | 1 | 1 | 1 |
| $f_1$ | $=$ | 0 | 0 | 1 | 1 | 0 | 0 | 1 | 1 |
| $f_0$ | $>$ | 0 | 1 | 0 | 1 | 0 | 1 | 0 | 1 |

Table 7.1: Specification of the test condition

$$D_{AU}(n) \;=\; D_{xor}(n) + \begin{cases} D_{CLA}(n) & \text{, carry look-ahead} \\ D_{RCA}(n) & \text{, ripple carry} \\ D_{CSA}(n) & \text{, conditional sum} \end{cases}$$

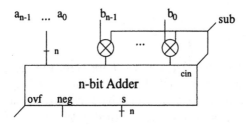

Figure 7.1: Circuit of the n-bit arithmetic unit $AU_n$

These are the only cost and run time formulae of the DLX design which the new adder designs impact directly. All the other formulae remain the same. Thus, these design changes only require minimal modification in the C program used to compute cost and cycle time of the DLX. The modified routines are listed in appendix B.7

## 7.2  Modifications of the Condition Test

Besides arithmetic operations, the ALU also compares inputs $a, b \in \{0, 1\}^n$. Thereby, the condition to be tested is coded by three bits f[2:0] as indicated in table 7.1.

In the original design (figure 7.2), the comparator circuit tests the difference $s = a - b$ instead of the inputs $a$ and $b$. The arithmetic unit AU provides the difference and the flag $less$ $(a < b)$ which corresponds to the AU flag $neg$. A zero tester tests $s$ for zero and thus generates the flag $equal$ $(a = b)$. When testing the result of the arithmetic unit, both circuits, AU and the comparator, lie on the time critical path, as the run time analysis of section 6.4.2 indicates.

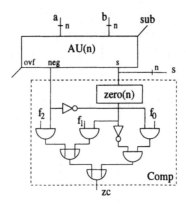

Figure 7.2: Arithmetic unit supplemented by the comparator circuit which tests the outputs of the AU

## 7.2.1 Testing ALU Inputs

Instead of testing the difference $s = a - b$ for zero, the comparator circuit can also test directly whether $a$ equals $b$, i.e., whether $a \otimes b$ equals zero. The zero tester and the arithmetic unit can then be traversed in parallel, and at most one of these circuits lies on the time critical path. Note, that the flag *less* is provided by the arithmetic unit.

As derived on page 83, the result $zc$ of the condition test can be expressed as

$$zc = (f_2 \, l \, \vee \, f_1 \, e) \, \vee \, (f_0 \, \bar{l}) \, \bar{e}.$$

For the inputs $a$ and $b$, the flags $l$ (less) and $e$ (equal) indicate whether $a < b$ or $a = b$, and the bits $f[2:0]$ code the condition to be tested as shown in table 7.1.

Since the arithmetic unit usually has a longer delay than the zero tester, we transform the computation of $zc$ as follows

$$zc = (f_2 \, l \, \vee \, f_1 \, e) \, \vee \, \bar{l} \, (f_0 \, \bar{e}).$$

Figure 7.3 depicts the modified circuit of the comparator and how to connect it to the arithmetic unit AU and to the logic unit. The cost of the new n-bit comparator circuit remains the same, namely

$$C_{Comp}(n) = C_{zero}(n) + 2 \cdot C_{inv} + 4 \cdot C_{and} + 2 \cdot C_{or}.$$

However, the modification impacts the delay $D_{ALU}(zc; n)$ of the ALU to compute the condition output $zc$:

$$
\begin{aligned}
D_{ALU}(zc; n) = \; & \max\{D_{AU}(n) + \max\{D_{and} + D_{or}, D_{inv} + D_{and}\}, \\
& D_{xor}(n) + D_{zero}(n) + \max\{D_{and} + D_{or}, D_{inv} + 2 \cdot D_{and}\}\} \\
& + D_{or}.
\end{aligned}
$$

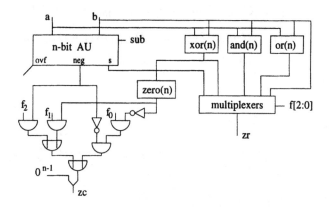

Figure 7.3: Circuit of the n-bit ALU, testing the inputs of the AU

Thus, with this modification of the comparator design, the cycle time can be up to $D_{zero}(32)$ delays faster at no additional cost. Once again, this modification only requires minimal changes in the program describing the cost and the cycle time of the DLX.

## 7.2.2  Two Cycle Condition Test

The most obvious way to speed up the cycle time is to split the time critical cycle into two (or more) cycles. In our case, the condition test is the critical cycle. Thus, we split the cycles Test and TestI from the pages 104 and 105 into two cycles each. During the first cycle, circuit AU computes the difference $s = a - b$ and generates the flag *less*. Together with the condition flags f[2:0], these data are stored in the additional (n+4)-bit register Cond. In the second cycle, the comparator reads these data, generates the result zc, and forwards it to the D bus. Register C then saves the result.

### Changes in the ALU

As far as the data paths are concerned, this modification only impacts the ALU (figure 7.4). Due to the additional register Cond, it now costs

$$C_{ALU}(n) = C_{AU}(n) + C_{Comp}(n) + C_{and}(n) + C_{or}(n)$$
$$+ C_{xor}(n) + 4 \cdot C_{mux}(n) + \mathbf{C_{ff}(n+4)}.$$

Compared to the original ALU design (figure 6.5), only the paths to output zc and to register Cond have changed. As before, $D_{AU}(32)$, $D_{Comp}(32)$ and $D_{ALUglue}$ denote the delay of the arithmetic unit, of the comparator and of the glue logic of the ALU environment. The modified paths through the ALU environment to register Cond and output ZC then have delays

$$D_{ALUenv}(S1, S2; Cond) = D_{AU}(32)$$

$$D_{ALU\,env}(Csig;\,Cond) = D_{ALU\,glue} + D_{AU}(32)$$

$$A_{ALU\,env}(Cond;\,ZC) = D_{Comp}(32) + D_{driv}(32)$$

$$D_{ALU\,env}(Csig;\,ZC) = D_{driv}(32).$$

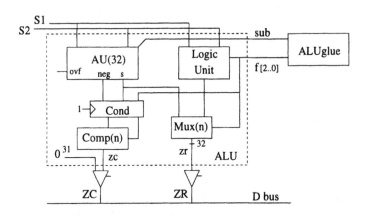

Figure 7.4: Modified ALU circuit and ALU environment

### Cycle Time of the Condition Test

In order to determine the cycle time of the two condition test cycles, we use a simplified view of the DLX data paths, as we did in section 6.4.2. Figure 7.5 only depicts those cycles in the DLX data paths which have changed. All the circuits shown there are governed by the control signals $Csig$. Thus, the cycles ending in register Cond require time

$$T_{ALU}(;\,Cond) = \max\{A_{CON}(Csig) + D_{ALU\,env}(Csig;\,Cond),$$
$$A_{RegS} + D_{ALU\,env}(S1,\,S2;\,Cond)\} + \Delta$$

and the cycles going through output ZC require time

$$T_{ALU}(ZC) = \max\{A_{CON}(Csig) + D_{ALU\,env}(Csig;\,ZC),$$
$$A_{ALU\,env}(Cond;\,ZC)\} + D_{Regw} + \Delta.$$

### Changes in the Control Unit

In contradiction to the design modifications we considered so far, this modification also impacts the control unit. In the FSD of the DLX (figure 6.18), we must replace the states **Test** and **TestI** by the sequences **Test1**, **Test2** and **TestI1**, **TestI2** as indicated in figure 7.6. Thus, the FSD comprises two more states than before and $k = 24$. However, $\zeta = \lceil \log k \rceil$ still equals 5.

Since we did not add new control signals and since the new states just take over
the functionality of the two state which they replace (table 7.2), the frequencies
of the active control signals remain the same. The transition from the first to
the second state of the condition test always occurs independent of any input
signals, i.e., the corresponding monomial is trivial and the new arcs have weight
1. Thus, the maximal fanin of the nodes in the FSD remains the same, but the
accumulated fanin $\text{fanin}_{sum}$ is increased by two. Altogether, only the parameters
$k = 24$ and $\text{fanin}_{sum} = 34$ of the control unit have changed.

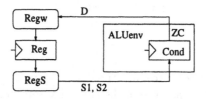

Figure 7.5: Simplified data paths for the test cycles

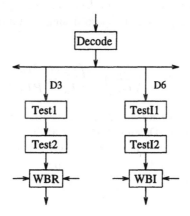

Figure 7.6: Modified section of the FSD

# 7.3   Evaluation of the Design Modifications

In the previous two sections, we introduced two modifications of the adder design
and of the implementation of the condition test. Strictly speaking, we suggested
to replace the carry look-ahead adder by a ripple carry or conditional sum adder,
to test the ALU inputs instead of their difference, or to perform the condition
test in two cycles. Altogether, that results in nine DLX design variants including
the original design. In addition, we can choose from four implementations of the

| State | Active Control Signals |
|-------|------------------------|
| Test | AS1doe, BS2doe, Cce, AluCDdoe, test |
| Test1 | AS1doe, BS2doe, test |
| Test2 | Cce, AluCDdoe |
| TestI | AS1doe, IRS2doe, Cce, AluCDdoe, test, Itype |
| TestI1 | AS1doe, IRS2doe, test, Itype |
| TestI2 | Cce, AluCDdoe |

Table 7.2: Active control signals of the removed states and of the new states in the FSD

control automaton. In order to limit the design variants, we assume for the rest of this chapter that that instance of the control automaton is used which proved to be most cost effective for the original DLX design (section 6.4.3), i.e., a Mealy implementation with states coded in binary.

Let A and B be two design variants with cost $C_A$ and $C_B$, and with cycle time $t_A$ and $t_B$. Let $CPI_A(Be)$ and $CPI_B(Be)$ indicate the CPI ratio of the variants A and B on a fixed workload $Be$. According to sections 2.2 and 2.3, the quality ratio $Q_q(A, B, Be)$ of the two design variants A and B can then be expressed as:

$$Q_q(A, B, Be) = \left(\frac{C_B}{C_A}\right)^q \cdot \left(\frac{t_B \cdot CPI_B(Be)}{t_A \cdot CPI_A(Be)}\right)^{(1-q)} .$$

The control automata of all six design variants with 1-cycle condition test are based on the original FSD, and thus they all have the same CPI ratio. The remaining three design variants have another CPI ratio. For comparisons within one of the two groups, the quality ratio only depends on the hardware parameters $C_A$, $C_B$, $t_A$ and $t_B$ and is independent of the workload. The quality ratio can then be expressed as

$$Q_q(A, B, ) = Q_q(A, B, Be) = \left(\frac{C_B}{C_A}\right)^q \cdot \left(\frac{t_B}{t_A}\right)^{(1-q)} .$$

However, design variants of different groups have different CPI ratios and can only be compared on the basis of a workload. Therefore, we first evaluate the design changes which do not impact the CPI value. Then, we introduce the benchmark workload and compare design variants with different CPI ratios.

## 7.3.1 Workload Independent Comparison

To compute the cost and cycle time of all nine design variants, a few new C routines are required; they are listed in appendix B.7. Table 7.3 lists the cost of the ALU and the cost and cycle time of the DLX for all nine design variants under

| Technology | | Motorola | | | Venus | | |
|---|---|---|---|---|---|---|---|
| | | Test Result | Test Inputs | 2 Cycle Test | Test Result | Test Inputs | 2 Cycle Test |
| RCA | $C_{ALU}$ | 1640 | 1640 | 1928 | 1964 | 1964 | 2396 |
| | $C_{DLX}$ | 11688 | 11688 | 11996 | 29109 | 29109 | 29561 |
| | $t_{DLX}$ | 164 | 155 | 151 | 91 | 86 | 85 |
| CLA | $C_{ALU}$ | 1858 | 1858 | 2146 | 2182 | 2182 | 2614 |
| | $C_{DLX}$ | 11906 | 11906 | 12214 | 29327 | 29327 | 29779 |
| | $t_{DLX}$ | 80 | 71 | 67 | 49 | 44 | 43 |
| CSA | $C_{ALU}$ | 2341 | 2341 | 2629 | 2665 | 2665 | 3097 |
| | $C_{DLX}$ | 12389 | 12389 | 12697 | 29810 | 29810 | 30262 |
| | $t_{DLX}$ | 50 | 49 | 49 | 41 | 36 | 35 |

Table 7.3: Cost and cycle time of the nine DLX variants

the assumption that the memory status time $d_{mstat}$ requires only 5 gate delays. Nevertheless, under Motorola technology, the computation of the memory status signals and the update of the control registers become the time critical path at least for the two fastest designs (CSA adder, testing inputs respectively with 2 cycle condition test).

These figures indicate that the fastest design variants in either group use a conditional sum adder (CSA). Compared to the original DLX design, the modifications reduce the run time between 10 and 39% and increase the cost by less than 7%. Thus, the modifications improve the price/performance ratio of the DLX up to 35%. Testing the inputs of the ALU instead of their difference reduced the cycle time (2 to 12%) at no additional cost and therefore always improves the quality of the design.

Switching from a ripple carry adder (RCA) to a conditional sum adder (CSA) reduces the cycle time by at least 55%, but the modification of the condition test improves the run time by 16.3%, at most. Thus, choosing the proper adder design has a bigger impact on the quality of the DLX than the modifications of the comparator design.

In order to perform more specific design evaluations, for any two design variants A and B of the same group, their equal quality parameter $EQ(A, B, Be)$ is required. This EQ value specifies the quality parameter $q$ for which design A and B have the same quality, i.e., $Q_q(A, B, Be) = 1$. For quality parameter $q < EQ(A, B, Be)$, the faster of the two designs is better, and for $q > EQ(A, B, Be)$, the cheaper of the two designs is better.

Since $EQ()$ is symmetrical in the first two arguments, we only compute the equal quality parameters for variants A and B, with A faster than B. The tables 7.4 and 7.5 list these equal quality parameters. The design variants $Vi$ are ordered by decreasing cycle time (and increasing cost). The values below the diagonal correspond to Motorola technology, the values above the diagonal

| Version | | | Venus Technology | | | | | |
|---|---|---|---|---|---|---|---|---|
| | Adder | Test | V1 | V2 | V3 | V4 | V5 | V6 |
| V1 | RCA | Result | | *1.000* | 0.988 | 0.990 | 0.971 | 0.975 |
| V2 | RCA | Inputs | 1.000 | | 0.987 | *0.989* | 0.969 | 0.973 |
| V3 | CLA | Result | 0.975 | 0.973 | | 1.000 | 0.916 | 0.950 |
| V4 | CLA | Inputs | 0.978 | *0.977* | 1.000 | | 0.812 | *0.925* |
| V5 | CSA | Result | 0.953 | 0.951 | 0.922 | 0.898 | | 1.000 |
| V6 | CSA | Inputs | 0.954 | 0.952 | 0.925 | *0.903* | 1.000 | |
| | | | Motorola Technology | | | | | |

Table 7.4: Break-even point $EQ(Vi, Vj)$ in the quality of the DLX design variants with 1-cycle condition test. The design variants are ordered by increasing cost and decreasing cycle time

| Version | | Venus Technology | | |
|---|---|---|---|---|
| | Adder | V7 | V8 | V9 |
| V7 | RCA | | 0.989 | 0.974 |
| V8 | CLA | 0.978 | | 0.928 |
| V9 | CSA | 0.952 | 0.890 | |
| | | Motorola Technology | | |

Table 7.5: Break-even point $EQ(Vi, Vj)$ in the quality of the DLX design variants with 2-cycle condition test

correspond to Venus technology.

The EQ values of design variants $V_i$, $V_j$ which test the inputs respectively the outputs of the AU and which are based on the same adder design equal 1, i.e., the faster of the two designs is always better. That is not surprising, because those pairs of designs have the same cost. The remaining EQ values are greater than 0.8, i.e., a slower but cheaper design variant will only be chosen if the cost of a design is far more important than its performance. Hence, under a realistic quality metric $Q_q$ with $q \in [0.2, 0.5]$ only the fastest variants are relevant, namely the variants with conditional sum adder which either test the ALU inputs (V6) or use a 2-cycle condition test (V9).

It is very surprising that the variants with conditional sum adder (CSA) win over those with carry look-ahead adder (CLA) so clearly, all the more so since the fixed-point adder in microcomputers is usually based on a CLA variant. However, our model ignores the fanout of gates. In the CLA adder, all gates have a small fanout, but in the CSA adder, the select signals of some multiplexers have to drive up to $n = 32$ gates. Thus, ignoring the fanout heavily favors the design variants with CSA adder over those with CLA adder. In the exercises

7.1 and 7.2 (section 7.4), we therefore discuss the impact of a fanout restriction
in great detail. Those analyses show that for a fanout of 2, i.e., an output can
drive at most two inputs, the CLA adder becomes cheaper *and faster* than the
CLS adder, and thus, the CLA adder becomes the superior design.

### The Local q-Range

For each design variant A of either group, we assign a range $Rl_A \subset [0,1]$ of
values $q$ such that any other variant B within the same group is of no higher
quality than variant A, i.e., for all $q \in Rl_A$, $Q_{q,A}(Be) \geq Q_{q,B}(Be)$. We call such
a range $Rl_A$ the *local q-range* of design A.

The q-ranges, some of which might be empty, can be derived as follows:

1. For each group of designs, we start with the fastest design. If there are
   several of those, we pick the cheapest of them. This design $A$ is the best
   of its group for $q = 0$ (run time only).

2. As already stated earlier, for an increasing quality parameter $q$, cost becomes more and run time becomes less important. Thus, we now look for
   a design $B$ of the same group as $A$ but cheaper than $A$ with a minimal EQ
   value $EQ(A, B,) < 1$. If there is no such design $B$, design $A$ is the best of
   its group independent of $q$ and its local q-range is $Rl_A = [0, 1]$.

   If such a $B$ exists, $Rl_A = [0, EQ(A, B,)]$, and for $q > EQ(A, B,)$, $B$ is
   better than $A$. We then look for a design $C$ cheaper than $B$ with minimal
   EQ value $EQ(B, C,) < 1$. If there is no such a $C$, $Rl_B = [EQ(A, B,), 1]$,
   otherwise $Rl_B = [EQ(A, B,), EQ(B, C,)]$ and we proceed in the same way
   till the interval $[0, 1]$ is covered completely.

For the nine DLX design variants V1 to V9, these ranges can easily be read
off from the tables 7.4 and 7.5, as we now demonstrate for the designs V1,..., V6
under Motorola technology. Figure 7.7 lists all non-empty q-ranges $Rl_A$ under
Motorola or under Venus technology.

Under Motorola technology, design variant $V6$ is the fastest of its group (table
7.3). Row V6 of table 7.4 lists the EQ values of design V6 with the other five
variants. The smallest of these values corresponds to variant V4. Thus, for
$q \in [0, 0.903[$, V6 is the best design, but for bigger $q$, design V4 is of higher
quality. Row V4 of the same table lists the EQ values of variant V4 with the
cheaper variants V1,..., V3. The smallest value belongs to variant V2. Thus,
variant V4 has the range $Rl_{V4} = [0.903, 0.977]$. V2 is the next relevant variant,
and it can not be replaced by variant V1 because $EQ(V1, V2) = 1.0$; thus,
$Rl_{V2} = [0.977, 1]$.

## 7.3.2   Workload Dependent Comparison

The modifications of the implementation of the condition test also impact the
FSD of the DLX. Consequently, these modifications can only be evaluated on
the basis of a fixed workload.

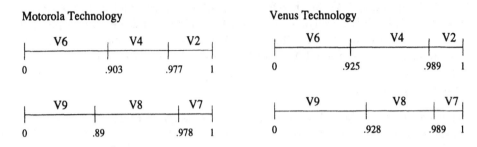

Figure 7.7: Segmentation of the interval of the quality parameter q. For $q$ in range $Rl_{Vi}$, design variant $Vi$ has a higher quality $Q_{q,Vi}$ than any other design variant of the same group.

## Benchmarks

Usually, a benchmark given in a high level programming language defines the workload. The programs first have to be complied into the machine language of DLX. Then, a dynamic run time analysis has to be performed. That yields statistics on the frequency of each machine instruction.

For our analyses, we use three benchmark kernels which Hennessy and Patterson also applied to the DLX architecture in [HP90]. So, we can use their statistics on the frequency of the DLX machine instructions. *GCC* and *TeX* are two C kernels. GCC is a Gnu C compiler applied to its own source code. In the TeX benchmark kernel, the TeX document-processing program is applied to a set of texts. The third kernel is the US Steel Cobol benchmark.

Table 7.6, which is taken from [HP90] (page C-5) lists the DLX instruction mix of all three benchmark kernels. The values do not add up to 100%, because the authors omitted instructions with a frequency of less than 1.5%. We therefore normalize these values to 100% (table 7.7).

The statistics indicate how often conditional branches do occur but not how often they are taken. However, as a rule of thumb [HP90] 53% of the branches are taken, at least for the C kernels. We use this additional information, in order to adapt the statistics to our needs (table 7.7). Besides the instruction counts, we also list the amount of pure CPU cycles and of memory accesses for each machine instruction.

The CPI value of machine instruction I is the number of its CPU cycles plus the product of memory accesses and cycles per access. With the figures of table 7.7, we now compute the CPI values of the DLX designs assuming that a memory operation on average requires WS wait states. Table 7.8 lists these CPI values.

## The Global q-Range

For any design variant $A$ with a non-empty local q-range $Rl_A$ we now determine a *global q-range* $R_A$. For a quality parameter $q \in R_A$, $A$ has no worse quality

| Instruction | GCC | TeX | US Steel |
|---|---|---|---|
| Control | | | |
| Branch | 19% | 7% | 16% |
| J | 2% | | 3% |
| JR | | | 2% |
| JAL, JALR | | | 2% |
| Arithmetic, logical | | | |
| ADDU, ADDUI | 17% | 20% | 27% |
| LHI | 2% | 10% | 3% |
| LI | 4% | 4% | 6% |
| SLL | 5% | 5% | 4% |
| SRA | 2% | | 2% |
| Test Set | 5% | 3% | 3% |
| AND, ANDI | 2% | | 3% |
| OR, ORI | | | 2% |
| Data transfer | | | |
| LW | 18% | 19% | 5% |
| SW | 10% | 12% | 5% |
| LBU | | 2% | |
| Total | 85% | 82% | 82% |

Table 7.6: Instruction mix of GCC, TeX and US Steel on the DLX architecture. Blank spaces in the table mean that this instruction or the group of instructions require less than 1.5% of all instructions in the benchmark. The same holds for instructions, not mentioned in the table. This table is adapted from Figure C.4 of [HP90], page C-5.

| Instruction | Instruction count [%] | | | CPU cycles | Memory accesses |
|---|---|---|---|---|---|
| | GCC | TeX | US Steel | | |
| Branch, taken | 12 | 5 | 10 | 3 | 1 |
| not taken | 10 | 4 | 9 | 2 | 1 |
| J, JR | 2 | 0 | 6 | 2 | 1 |
| JAL, JALR | 0 | 0 | 3 | 3 | 1 |
| ALU | 29 | 41 | 49 | 3 | 1 |
| Test set | 6 | 4 | 4 | $3^1$ | 1 |
| Shift | 8 | 6 | 7 | 3 | 1 |
| Load | 21 | 25 | 6 | 4 | 2 |
| Store | 12 | 15 | 6 | 3 | 2 |

Table 7.7: Instruction mix of GCC, TeX and US Steel on the DLX architecture, normalized and rounded. The last two columns list the amount of CPU cycles and memory accesses required per machine instruction.

---

[1] this cycle count equals 4 for DLX variants with 2-cycle condition test.

| | 1-cycle Condition Test | 2-cycle Condition Test |
|---|---|---|
| Gcc | $4.42 + 1.33 \cdot WS$ | $4.48 + 1.33 \cdot WS$ |
| TeX | $4.61 + 1.4 \cdot WS$ | $4.65 + 1.4 \cdot WS$ |
| US Steel | $4.03 + 1.12 \cdot WS$ | $4.07 + 1.12 \cdot WS$ |

Table 7.8: Average CPI value of the DLX at WS wait states per memory access (on average)

than any of the other designs considered, i.e., $Q_{q,A}(Be) \geq Q_{B,q}(Be)$. The global q-range must be a subset of the local q-range for any design A:

$$R_A \subset Rl_A.$$

The global q-ranges can be derived as follows:

1. *Sorting Designs.* The designs with non-empty local q-ranges are ordered by decreasing cost. Since for an increasing quality parameter $q$, cost becomes more important, an expensive design if it is at all profitable must be so for some small $q$.

2. *Eliminating Designs.* If a design $B$ is more expensive but not faster than a design $A$, design $B$ is not profitable and can be dropped.

3. *Computing q-Ranges.* Let $A - B - C$ be a portion of the sorted list of designs left after the previous elimination. $A$ is the most expensive and

$C$ the cheapest of the three designs. Hence, for these three designs, the range $[0:1]$ of quality parameters can be partitioned as depicted in figure 7.8. Design $B$ is better than design $A$, if the quality parameter is at least $EQ(A, B, Be)$, i.e., $q \in [EQ(A, B, Be), 1] =: I_1$. On the other hand, design $B$ is better than design $C$, if the quality parameter is at most $EQ(B, C, Be)$, i.e., $q \in [0, EQ(B, C, Be)] =: I_2$. If the global q-range of the designs $A$ and $C$ are not empty, the q-range of design variant $B$ must be the intersection of the two intervals $I_1$ and $I_2$:

$$R_B \;=\; Rl_B \cap I_1 \cap I_2 \;=\; Rl_B \cap [EQ(A, B, Be), 1] \cap [0, EQ(B, C, Be)].$$

In the case that $B$ and $A$ (respectively $C$) belong to the same group of designs, i.e., they have the same CPI ration, the intersection with interval $I_1$ (respectively $I_2$) can be dropped.

In order to compute the q-ranges, we start with the most expensive design. At first, we only know the local q-ranges. Thus, some designs may drop out, and the computation of the q-ranges must start over again. Since in every iteration, the number of designs decreases, the process finally terminates.

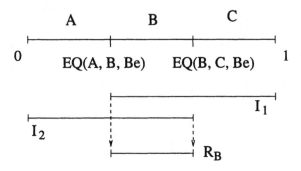

Figure 7.8: The global q-range of design $B$

**Comparison under Motorola Technology**

Figure 7.7 lists all design variants with non-empty q-range as well as their q-range. Ordering these six designs by decreasing cost (table 7.3) yields the following list

$$V9 \;-\; V6 \;-\; V8 \;-\; V7 \;-\; V4 \;-\; V2.$$

In this case, the designs $V9$ and $V6$ have the same cycle time, but $V9$ with its two cycle condition test is more expensive and has a larger CPI ratio than V6 which tests the AU inputs within one step. Thus, under the current workload,

$V6$ is always better than $V9$. The same arguments hold for the designs $V7$ and $V4$. After dropping designs $V9$ and $V7$, the list looks like

$$V6 - V8 - V4 - V2.$$

Thus, from the design variants with two cycle condition test, only design $V8$ (CLA adder) is left with a local q-range of $Rl_{V8} = [0.89, 0.978]$. Figures 7.9 and 7.10 depict the break-even points $EQ(V6, V8, Be)$ and $EQ(V4, V8, Be)$ as a function of the average number of wait states per memory access. The EQ value of the designs V6 and V8 lies above 0.95 and the EQ value of V4 and V8 lies below 0.7. Thus, the global q-range of $V8$ is empty:

$$R_{V8} = [0.89, 0.978] \cap [0.95, 1] \cap [0, 0.7] = \emptyset.$$

That rules out the variant V8 and the two-cycle condition test as well. The decrease in the cycle time is not big enough to compensate for the cost increase and for the additional machine cycles. For the designs $V6$, $V4$ and $V2$, the global and the local q-ranges are the same.

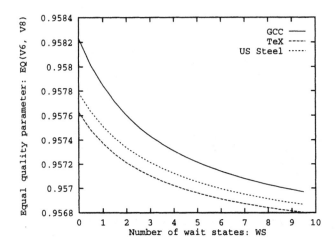

Figure 7.9: Equal quality parameter $EQ(V6, V8, Be)$ as a function of the average number WS of wait states per memory access (Motorola technology)

### Comparison under Venus Technology

Under Venus technology, the same six designs have a non-empty local q-range (figure 7.7) and ordering them by decreasing cost (table 7.3) results in the same list as before

$$V9 - V6 - V8 - V7 - V4 - V2.$$

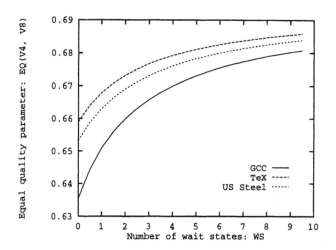

Figure 7.10: Equal quality parameter $EQ(V4, V8, Be)$ as a function of the average number WS of wait states per memory access (Motorola technology)

Design $V7$ is still more expensive and slower than $V4$, but for $V9$ that is no longer the case. Thus, the reduced list looks like

$$V9 \; - \; V6 \; - \; V8 \; - \; V4 \; - \; V2.$$

Figure 7.11 depicts the equal quality value $EQ(V4, V8, Be)$ of the two designs as a function of the average number of wait states per memory access under all three benchmarks. For quality parameters $q$ greater than $EQ(V4, V8, Be)$, design variant V4 is better and for the others, V8 is better. The EQ value is less than 0.61, thus, the global q-range of $V8$ obeys

$$R_{V8} \subset [0.89, 0.978] \cap [0, 0.61] \; = \; \emptyset.$$

That rules out variant V8 with 2-cycle condition test and CLA adder.

Figure 7.12 depicts the equal quality value $EQ(V6, V9, Be)$ of the two designs as a function of the average number of wait states per memory access under all three benchmarks. Depending on the workload and on the speed of the main memory, the break-even point of these two designs lies between $q = 0.494$ and $q = 0.65$. Thus, under a realistic quality model $Q_q$ with $q \in [0.2 : 0.5]$, the variant with 2-cycle condition test (V9) always wins over the one with 1-cycle test (V6). Only on a GCC like workload and in combination with very fast main memory (no wait states), variant V6 can compete with V9 on the basis of cost/performance.

Once again, under a realistic quality metric $Q_q$ with $q \in [0.2, 0.5]$ and without a fanout restriction, only two variants are realistic, namely the variants with

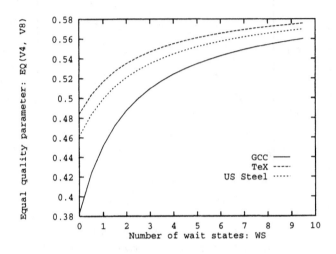

Figure 7.11: Equal quality parameter $EQ(V4, V8, Be)$ as a function of the average number WS of wait states per memory access (Venus technology)

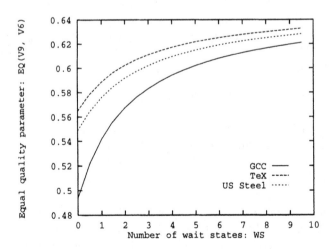

Figure 7.12: Equal quality parameter $EQ(V6, V9, Be)$ as a function of the average number WS of wait states per memory access (Venus technology)

conditional sum adder which either test the ALU inputs (V6) or use a 2-cycle condition test (V9). Variant V6 is better under Motorola technology and variant V9 under Venus technology.

## 7.4   Exercises

7.1 In exercise 3.6 of section 3.9, we derived variants $FRCA_{n,f}$, $FCLA_{n,f}$ and $FCSA_{n,f}$ of the original adder designs which satisfy a fanout restriction of $f$.

    1. Perform the trade-off analysis for a fanout restriction of $f = 2$ ($f = 4$), based on the delay and the cost of the optimized adders described above.

    2. How does the fanout restriction impact the trade-offs?

7.2 Under Motorola technology, we optimized these adders even further and got variants $FMRCA_{n,f}$, $FMCLA_{n,f}$ and $FMCSA_{n,f}$ (section 3.9, exercise 3.6).

    1. Perform the trade-off analysis for a fanout restriction of $f = 2$ ($f = 4$), based on those Motorola specific adder variants (under Motorola technology only).

    2. How does the Motorola specific optimization impact the trade-offs?

    3. Compare the results with those of the previous exercise gained for the Venus technology.

# Chapter 8

# Interrupt

Interrupts are events which may change the sequential flow of control of a program by means other than a branch instruction. They are triggered by the activation of *event signals*. We will denote these signals by $evn[1]$, $evn[2]$, .... Here, we will treat in detail the following interrupts:

- reset

- illegal instruction

- misaligned memory access

- page fault during instruction fetch

- page fault during load or store

- trap (a special machine instruction which causes an interrupt)

- overflow

- a number $q$ of additional interrupts $ex_j, j \in \{1, \ldots, q\}$ produced by external I/O-devices, where $q = 32 - 7 = 25$.

Some interrupts change the flow of control whenever they occur, others are *maskable*, i.e., they can be ignored under software control.

Programs executed as a consequence of interrupts are called *interrupt service routines* (ISRs). Executing such a routine is also called servicing an interrupt. The service starts with the fetch cycle of the first instruction of ISR and ends with the last cycle of the instruction used to leave the ISR.

We will use a single interrupt service routine, and by *SISR* we denote its start address. Once the routine is activated, the machine determines the cause of the interrupt and jumps to a section of the ISR specific to the detected cause. After servicing the interrupt, the machine leaves the ISR.

| Name | Symbol | External | Maskable | Resume |
|------|--------|----------|----------|--------|
| reset | reset | yes | no | abort |
| illegal instruction | ill | no | no | abort |
| misaligned access | misa | no | no | abort |
| page fault on fetch | pff | no | no | repeat |
| page fault on load/store | pfls | no | no | repeat |
| trap | trap | no | no | continue |
| arithmetic overflow | ovf | no | yes | continue |
| external I/O | $ex_j$ | yes | yes | continue |

Table 8.1: Symbols and type of the different interrupts

**Notation:** For all $i$ we denote by *interrupt* $i$ the interrupt corresponding to event signal $evn[i]$. $ISR(i)$ indicates the *process* of running the interrupt service routine in order to service interrupt $i$.

Before we can even think of designing hardware that supports interrupts, we have to answer the following questions for each type of interrupt:

- Is the event signal generated by the CPU or is it external?

- During what kind of instruction and during which cycle of execution of that instruction can the event signal become active?

- How long does the signal stay active and how is it cleared?

- Which interrupts are maskable?

- What is the *priority* of the interrupt? These priorities regulate, under which circumstances interrupt service routines themselves can be interrupted and which interrupt has the right of way in case several event signals become active simultaneously.

- Where do we resume the execution of the interrupted program after completion of the interrupt service routine? The three possible alternatives are:

  - abort the whole program

  - repeat the interrupted instruction $I$

  - continue with the instruction $I'$, which follows $I$ in the noninterrupted execution of the program

Tables 8.1 and 8.2 answers most of these questions for the above interrupts. The number of an interrupt also indicates its priority level. As usual, high priority is denoted by *small* numbers.

| Symbol | Priority, $evn[]$ | Instruction | Machine Cycle |
|--------|-------------------|-------------|---------------|
| reset | 0 | any | any |
| ill | 1 | illegal | decode |
| misa | 2 | any | instruction fetch |
| | | load, store | data read/write |
| pff | 3 | any | instruction fetch |
| pfls | 4 | load, store | data read/write |
| trap | 5 | trap | decode |
| ovf | 6 | add, addi, sub, subi | execute |
| $ex_j$ | $6 + j$ | any | any |

Table 8.2: Priority and event number of the interrupts, and the machine instruction and machine cycle during which they can occur

The external event signals are provided by external inputs, which basically behave like the outputs of a register. Besides the powerup signal *pup*, we will need 26 such inputs, namely $evn[0] = reset$, $evn[7] = ex_1$, $evn[8] = ex_2, \ldots$. For the external interrupts, we make the following

**Interrupt Convention:** An active event line $evn[j]$ with $j \geq 7$ is only turned off when the interrupt $j$ is serviced, i.e., the device which activated the event line turns the signal off before the machine leaves ISR(j) or ISR(i) with $i < j$ and interrupt $i$ aborts the whole program.

That specifies how long external event signals stay active. Internal interrupts are only related to the current instruction, and so the event signal only stays active for one cycle, i.e., the cycle during which the signal is generated.

In the next sections, we describe how to support interrupts, extend the DLX design and evaluate the corresponding design changes.

# 8.1 Supporting Interrupts

In this section, we describe hardware *and software* for supporting the interrupts introduced in the previous section in the following order:

1. special registers and special instructions

2. execution of interrupt service routines

3. extensions of the data paths

4. control

## 8.1.1  Special Purpose Registers and Special Instructions

**Notation:**  An external interrupt $j$ is called *pending* if it has activated its event signal, and if since then neither interrupt $j$ nor an interrupt which aborts the whole program and which has a higher priority than interrupt $j$ is serviced.

An internal interrupt $j$ is called *pending* if it had activated its event line, and at that time interrupt $j$ was not masked, and if since then neither interrupt $j$ nor an interrupt which aborts the whole program and which has a higher priority than interrupt $j$ is serviced.

Pending can not be defined in the same way for internal and external interrupts, because masking has a fundamentally different impact on both groups of interrupts. If an internal interrupt occurs while it is masked (in our case only for overflow), it is ignored altogether. If an I/O interrupt is masked, it is temporarily ignored and should receive service once it is unmasked.

For the support of interrupts we introduce 6 new registers:

1. the status register $SR$. For now, we only use bits SR[31:6] of this register for masking interrupts $31, \ldots, 6$. For $j \in \{6, \ldots, 31\}$, SR[j]=1 enables interrupt $j$, SR[j]=0 masks it.

2. the cause register $CA$. For all $i \leq 31$ we use bit CA[i] for recording, that interrupt $i$ is pending.

3. the exception program counter $EPC$. On a jump to the ISR, the address $I$ of the interrupted instruction is saved in the register EPC. When we return from the interrupt service routine, EPC holds the return address.

4. the exception status register $ESR$. On a jump to the ISR, register SR is saved in ESR. When we return from the service of an interrupt, ESR holds the old content of SR.

5. the exception cause register $ECA$. On a jump to the ISR, register ECA takes over the masked content MCA of the cause register, i.e.,

$$ECA[i] = MCA := \left\{ \begin{array}{ll} CA[i] & , i \leq 6 \\ SR[i] \wedge CA[i] & , \text{otherwise} \end{array} \right.$$

and register CA is cleared. Note, the event signal of the internal interrupt overflow was already combined with its mask bit SR[6] before it got clocked into the cause register CA. There is no need for masking it twice.

6. the exception memory address register $EMAR$. On a jump to the ISR, the value of the memory address register MAR is saved in EMAR. On an interrupt caused by a failure in a memory data access, EMAR provides the address of the faulty memory access to the ISR.

   With self modifying code, the faulty memory address could also be reconstructed from the instruction word, but we will not take that route.

| Register | SR | ESR | EPC | ECA | EMAR |
|----------|-----|------|------|------|------|
| IR[10:6] | hx00 | hx01 | hx02 | hx03 | hx04 |

Table 8.3: Addresses of special purpose registers (hexadecimal)

| IR[31..26] | Mnemonic | Effect |
|------------|----------|--------|
| hx3e | trap | evn[5] = 1;  EPC = PC;  PC = SISR; ESR = SR;  ECA = MCA;  SR = 0 clear CA but catch new interrupt events |
| hx3f | rfe | SR = ESR;  PC = EPC |

Table 8.4: Coding and impact of the two interrupt instructions. SISR is the start address of the ISR, MCA indicates the masked content of the cause register.

During nested interrupts, the exception registers EPC, ESR, ECA and EMAR must be saved on an *interrupt stack*.

The cause register CA belongs to the control unit, and it is not visible to the programmer. The other new registers, called *special purpose registers*, can be explicitly accessed with the R-type data transfer instructions **movs2i** and **movi2s** (see table 6.2). The addresses of the special registers are specified by field *SA* of the instruction word according to table 8.3. In order to save a special purpose register on the interrupt stack, its content first must be moved to a general purpose register.

Two new J-type instructions, **trap** and **rfe** are needed. Their effect is described in table 8.4. The trap is a user requested jump to the ISR. The type of the trap can be specified — in a somewhat dirty way — by the 26-bit constant. The ISR can then use the saved exception PC in order to load the trap instruction into a register and to extract the constant. The machine uses the instruction **rfe** (return from exception) to leave the ISR. The use of the instructions is explained in the following section.

## 8.1.2 Execution of an Interrupt Service Routine

Once an active event line has been detected the following steps are executed:

1. Let $I$ be the interrupted instruction. It depends on the way in which the program is to be continued after the interrupt, whether instruction $I$ is completed, or whether the state of the machine before instruction $I$ got started is restored. The possible combinations are:

   - On an interrupt of type "continue", the current instruction $I$ is completed.

- On an interrupt of type "repeat", the state of the machine before instruction $I$ is restored.

- On an interrupt of type "abort", it does not matter.

2. The next few steps save the status of the DLX data paths on the interrupt stack. We denote this sequence by *Save Status*.

   (a) Registers *PC, SR, MCA* and *MAR* are saved into their counterparts *EPC, ESR, ECA, EMAR*. All maskable interrupts are masked, i.e., the masks in the status register are cleared. The start address *SISR* of the interrupt service routine is loaded into the *PC*.

      If we could guarantee, that no other interrupt of higher priority occurs for a few instructions, then we only would have to jump to *SISR* under hardware control, and we could do the rest with the first few instructions of the interrupt service routine. However, if we get caught during this activity by another interrupt, then things become very messy very quickly. Therefore, all these activities are performed in an *atomic*, i.e., non interruptable way under hardware control. This also explains, why the effect of the *trap* instruction in table 8.4 consists in more than just a jump to *SISR* and setting a bit in the cause register.

   (b) The interrupt service routine saves *EPC, ESR, ECA* and *EMAR* on the interrupt stack in main memory. That requires the use of a general purpose register which, at forehand, is also saved on the interrupt stack.

   (c) Let *eca* be the content of register ECA saved in the previous step. If $eca[0] = 0$, the interrupt service routine increments the interrupt stack pointer by 35. Thus, providing enough space to save all nontrivial general purpose registers and the exception registers EPC, ESR, ECA and EMAR.

      On reset ($eca[0] = 1$), the interrupt service routine does not create a new frame on the interrupt stack but clears the interrupt stack, and directly jumps to the section of ISR specific to reset.

   (d) The interrupt service routine determines the number

      $$l = \min\{j \mid eca[j] = 1\}$$

      of the pending interrupt with highest priority and unmasks all interrupts with higher priority.

3. From now, interrupt $l$ is serviced and the ISR can be interrupted by interrupts with higher priority. The ISR jumps to code specific to interrupt $l$ and saves those general purpose registers which it will use later on but which it has not saved yet. In step 2c, the ISR already allocated enough space on the interrupt stack for the content of these registers.

4. Before the execution of the **rfe** instruction, the general purpose registers are restored from interrupt stack and all maskable interrupts are masked[1].

5. The following two steps, called *Restore Status*, restore the status of the DLX data paths.

   (a) The return address and the old content of SR are restored from the interrupt stack in EPC respectively in ESR. The interrupt stack pointer is decremented by 35 and the general purpose register used to load EPC and ESR is restored.

   On a reset ($eca[0] = 1$), the registers are initialized instead of restored. That also includes the interrupt stack pointer.

   (b) In the **rfe** instruction, PC and SR are replaced by EPC and ESR in an atomic way.

In step 2c), the ISR clears the interrupt stack on reset but not on other aborting interrupts like illegal instruction or misaligned memory access. That allows the system to report an error and to shut down gracefully on those aborting interrupts.

A correct servicing of the interrupts can only be guaranteed if the steps 2 (Save Status) and 5 (Restore Status) are not interrupted except by Reset. Thus a careful design of the ISR is necessary so that the sequences Save Status and Restore Status become *protected regions*.

We will hold both the interrupt stack and the ISR on *permanent* memory pages, i.e., pages which are never swapped out of main memory. This guarantees, that the ISR is never interrupted by page faults. Some convention like this is necessary: imagine that *no* part of the ISR is in main memory. Then, any interrupt will result in a page fault and this page fault cannot be serviced without producing another page fault, etc.

It is common practice to reserve a general purpose register for the stack pointer of the interrupt stack. Register 31 is already reserved for the return address on the two jump and link instructions **jalr** and **jal**. We therefore use register 30.

It is clear, that servicing interrupts requires *both* hardware support and a *software protocol* to be obeyed. The design of protocols of any kind is one of the most error prone activities in computer science. Thus, it is no wonder that this part of machine design is acknowledged to be particularly hard ([HP90], page 214). The best known defense against errors in the design of protocols is to attempt a correctness proof. We will take this route here. It should however be noted that this route is *not fail safe*, because correctness proofs for protocols are the most error prone proofs known in computer science.

In section 8.3, we will formalize what is meant by the correctness of the hardware and software supporting interrupts, and we will prove the correctness of the DLX design with interrupt handling given next.

---

[1] These instructions can possibly be interrupted by a maskable interrupt!

### 8.1.3    Extensions of the Data Paths

In order to execute the new instructions and to monitor the activity of event lines we have to adapt the data paths. That basically concerns the environments of the PC, of the main memory, of the special purpose registers and of the interrupt logic as well as the powerup mechanism.

#### The Powerup Mechanism

On powerup ($pup = 1$), the control automaton initializes its state register with the initial state $z_0$ and its output register, if there is any, with the corresponding output vector. This remains the same.

In contrast to the previous DLX design, an active powerup signal $pup$ will not directly pull the PC down to zero (i.e., the start address of the initial program). Instead, it activates the event signal of the hardware reset. Thus, the ISR of the reset interrupt initializes the registers of the data paths and the RAMs on reset and on powerup.

#### The PC-Environment

Due to the new powerup mechanism, the control circuit PCC (section 6.2.5) is not necessary any longer. The clock signal $PCce$ is now generated directly by the control automaton. The multiplexer controlled by the standard control signal $PCsel$ permits to load the start address $SISR$ of the interrupt service routine into the PC (figure 8.1). Cost and delays of the PC environment remain the same:

$$
\begin{aligned}
C_{PCenv}(32) &= 2 \cdot C_{driv}(32) + C_{ff}(32) + C_{mux}(32) \\
D_{PCenv}(Sbus; 32) &= D_{driv}(32) \\
D_{PCenv}(D; 32) &= D_{mux}(32).
\end{aligned}
$$

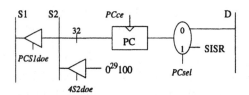

Figure 8.1: PC-environment. SISR indicates the start address of the ISR.

#### The Main Memory Environment

The main memory environment causes an interrupt on page fault and on a misaligned access. To protect the memory content and to provide enough time to respond to a memory problem, we make the following assumption:

On a page fault, the memory never updates its content, instead it activates its output flag *busy*. On misaligned memory accesses, the memory control circuit MC of section 6.2.6 generates no bank write signal. Thus, the memory content is also safe on a misaligned access.

The memory environment processes the busy signal of main memory and generates signal *not ready (nready)*

$$nready \ = \ busy \ \lor \ mis$$

which indicates that the memory access did not finish yet.

The main memory environment requires an additional OR gate in order to generate signal *nready*. Its cost therefore runs at

$$C_{Menv} \ = \ 2 \cdot C_{ff}(32) + 2 \cdot C_{driv}(32) + C_{mux}(32) + \mathbf{C}_{or}.$$

In sections 6.4.2 and 6.3.2, it turned out that the memory status signals and signal *mis* are valid $A_{Menv}(mstat)$ respectively $A_{CON}(mis)$ delays after the start of the cycle. These formulae still hold for the DLX with interrupt handling. Thus, signal *nready* is valid after $A_{Menv}(nready)$ gate delays, with

$$A_{Menv}(nready) \ = \ \max\{A_{CON}(mis), \ A_{Menv}(mstat)\} + D_{or}.$$

**Special Purpose Register Environment**

Two of the new registers are accessed in every cycle: the cause register is written in order to catch the event signals, and the mask register (SR) is read in order to enable or disable the propagation of these interrupts. Thus, the new registers cannot be realized as a single register file.

Since the cause register CA belongs to the control, for reason which will become clear in section 8.1.4, we treat the cause environment together with the control unit in the next section. The other new registers, called *special purpose registers* (SPR), are only accessed in the interrupt service routine, on jumps to ISR (JISR) and during return from ISR (RFE) as described in section 8.1.2. Table 8.5 summarized these actions. Since all registers must be read, we provide a tristate driver to the D bus for each of them. The registers ECA and EMAR get their inputs exclusively from signal MCA (masked content of CA) respectively from register MAR. Their inputs are therefore directly connected to that source. Since the other registers have at least two different sources, we connect their data inputs to the D bus. Figure 8.2 depicts the environment of the special purpose registers and how to connect it to the data paths.

So far, only the bits 6 to 31 of SR store the interrupt masks; the other six bits are unused. From now on, we use bit SR[0] to save the overflow flag *ovf* of the ALU. Thus, the programmer can access this bit directly and not only though interrupts. That is essential in order to provide efficient software routines for 64-bit arithmetic or for floating-point arithmetic.

Consequently, the content of the register SR can be changed in one of the following three ways:

|       | SR            | ESR       | EPC       | ECA       | EMAR      |
|-------|---------------|-----------|-----------|-----------|-----------|
| JISR  | move to ESR;  | take over | take over | take over | take over |
|       | clear         | SR        | PC        | MCA       | MAR       |
| RFE   | take over     | moved to  | moved to  | —         | —         |
|       | ESR           | SR        | PC        |           |           |
| ISR   | read masks;   | push;     | push;     | push      | push      |
|       | change masks  | restore   | restore   |           |           |

Table 8.5: Actions performed on the special purpose registers. Push indicates that the content of the register is moved to the interrupt stack. MCA denotes the masked content of the cause register.

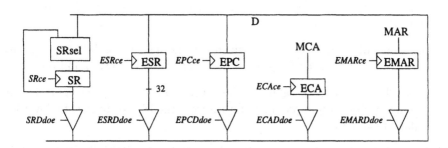

Figure 8.2: Special purpose register environment. MCA, the masked content of CA, is provided by the control unit.

- Update bit SR[0] = $ovf$ during the compute cycles of the instructions **add**, **addI**, **sub** and **subI**. The mask bits remain the same.

- Clear the whole register SR as a consequence of the interrupt handling.

- Load the whole register with the data provided on the D bus.

We use two multiplexers to select the proper input of register SR (figure 8.3). The control unit governs these multiplexers via two additional select signals $SRsl1$ and $SRsl2$.

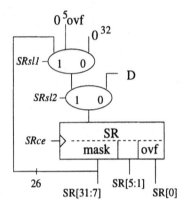

Figure 8.3: Circuit SRsel. Signal ovf is the overflow signal of the ALU

For simplicity, we make every register 32 bits wide, although a few bits may not be used. Thus, the environment of the special purpose registers has cost

$$C_{SPRenv} = 5 \cdot C_{ff}(32) + 5 \cdot C_{driv}(32) + 2 \cdot C_{mux}(32).$$

For every special purpose register $X \in \{EPC, SR, ESR, ECA, EMAR\}$ we select the clock enable signal $Xce$ and the output enable signal $XDdoe$ of the driver to the D bus from one of two sources:

- the output of a 3-bit decoder, which decodes instruction bits IR[8:6]. This source is used during the special move instructions **movi2s** and **movs2i**.

- signals $Xice$ and $XDioe$ which are generated directly by the control. This source is used to jump to or to return from the ISR, or to save the overflow flag of the ALU.

The control signals $spr$ (special read) and $spw$ (special write) indicate the special move instructions. These two signals select between the two sources described above. The resulting circuit SPRCon, which is part of the control unit, is shown in figure 8.4. Its cost runs at

$$C_{SPRCon} = C_{dec}(3) + 10 \cdot C_{and} + 10 \cdot C_{or}.$$

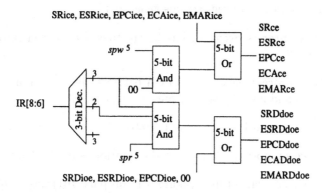

Figure 8.4: Selection of the clock and enable signals of the special purpose register environment

Let $A_{CON}(Csig)$ be the delay, which the control automaton requires in order to generate the control signals $spr$, $spw$, $XDioe$ and $Xice$ after the start of a new cycle. Then, the output signals of circuit SPRCon are valid $A_{SPRCon}$ gate delays after the start of the cycle, with

$$A_{SPRCon} = \max\{A_{CON}(Csig), D_{dec}(3)\} + D_{and} + D_{or}.$$

The following delays of the SPR environment are important for the cycle time of the DLX machine. After the start of a new cycle, the SPR environment requires $A_{SPR}(; D)$ gate delays to forward data to the D bus. Moreover, it takes $D_{SPR}(D;)$ gate delays, to forward data from the D bus to the special purpose registers, but it takes no time ($D_{SPR}(MCA) = 0$) to forward the masked cause register to register ECA.

$$A_{SPR}(; D) = A_{SPRCon} + D_{driv}(32)$$
$$D_{SPR}(D;) = D_{mux}(32)$$
$$D_{SPR}(MCA) = 0.$$

The D bus connects the environments but is not part of them. Thus, only the cycle from register SR through circuit SRsel lies completely within the environment of the special purpose registers. This cycle has the cycle time:

$$T_{SRsel}(SR; SR) = \max\{A_{CON}(Csig) + 2 \cdot D_{mux}(32), A_{SPRCon}\} + \Delta.$$

### 8.1.4  Control

In order to prepare the control unit for interrupts, we have to deal with the following issues:

1. How are the internal event signals generated?

2. How do we process event signals before they are clocked into the cause register CA?

3. How do we process the outputs of CA before we feed them into the control automaton and into the special purpose register ECA?

4. How do we extend the control automaton?

5. How do we support the hope that the solution is correct? (section 8.3)

Thus, the original control unit (figure 6.17 on page 101) has to be extended by circuit *SPRCon* of figure 8.4 and by the environment of the cause register CA. Due to the modified powerup mechanism, the control circuit PCC which governed the PC environment can be dropped. Circuit *GenEvn* of the cause environment generates the internal event signals, and processes all event signals before they are clocked into the cause register. An additional circuit *CaPro* (cause processing) processes the content of the cause register, feeds some of its results into the next state circuit of the control automaton, and forwards the other results to the special purpose register ECA. Figure 8.5 depicts the structure of the extended control unit on the instance of a Moore implementation of the control automaton with binary coded states.

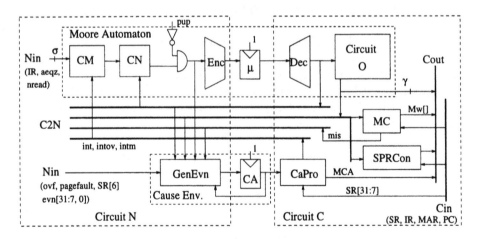

Figure 8.5: DLX control unit, using a Moore implementation with binary coded states

**Generation of Internal Event Signals**

There are 6 internal event signals: *illegal instruction (ill)*, *misalignment (misa)*, *page fault on fetch (pff)*, *page fault on load/store (pfls)*, *trap* and *overflow*.

**Illegal Instruction** For the generation of *illegal instruction*, we exploit the fact that the DLX machine language provides no explicit NOP instruction. Thus, after the decode state, any legal instruction switches to a state different from fetch and decode. However, in the Mealy and Moore implementations of the control automaton (chapter 4) the next state circuit also generates value 0 (fetch) if the next state is undefined. On an illegal instruction, the control automata therefore switches directly from decode to fetch. Testing for this special transition provides a simple way to check for an illegal instruction.

Figure 8.6 lists the hardware extensions for all four implementations of the control automaton. Only the Moore implementation with states coded in binary requires an additional $(k-1)$-bit zero tester. Thus, cost and delay of circuit *ILL* which generates event signal *ill* run at

$$C_{ILL} = \begin{cases} C_{zero}(k-1) + C_{and} & \text{for Moore with binary coding} \\ C_{and} & \text{otherwise} \end{cases}$$

$$A_{ILL} = D_{and} + \begin{cases} A_{NSE} - D_{enc}(\zeta) + D_{zero}(k-1) & \text{, Moore, binary} \\ A_{uNSE} & \text{, Moore, unary} \\ A_{NSD} - D_{enc}(\zeta) + D_{zero}(k-1) & \text{, Mealy, binary} \\ A_{uNSD} & \text{, Mealy, unary} \end{cases}$$

In the Moore implementation with states coded in binary, we use a $(k-1)$-bit zero tester, because that is faster than the combination of a $\zeta$-bit encoder and a $\zeta$-bit zero tester. However, the zero tester is still slower than the encoder. With this instance of the control automaton, it therefore takes longer to generate event signal $evn[1]$ than to compute the next state of the control. Thus, this cycle becomes a possible candidate for the critical path of the control unit. On the other implementations of the control automaton, signal *ill* requires no more time than updating the state of the automaton.

According to the specification of the control circuits N and C in section 2.4.5, all outputs of circuit N feed into registers of the control unit. Since circuit ILL receives inputs from N, circuit ILL and the cause register must be part of the control unit as well.

**Memory Problems** The signals *mis* and *pagefault* are directly provided by the memory environment, but the memory is not only enabled on read and write cycles. The control unit therefore generates two control signals *fetch* and *ls* which indicate a memory access on instruction fetch respectively on load or store. Then, the three event signals *misa, pff* and *pfls* which indicate memory problems can be derived as:

$$\begin{aligned} misa &= evn[2] &= mis \wedge (fetch \vee ls) \\ pff &= evn[3] &= pagefault \wedge fetch \\ pfls &= evn[4] &= pagefault \wedge ls \end{aligned}$$

**Trap and Overflow** A trap will get its own state in the control automaton, which activates control signal $evn[5] = trap$.

**Moore, binary**

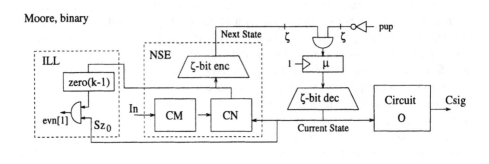

**Moore, unary**

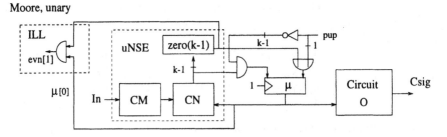

**Mealy, binary**

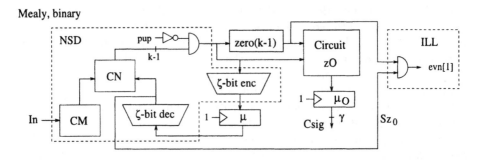

**Mealy, unary**

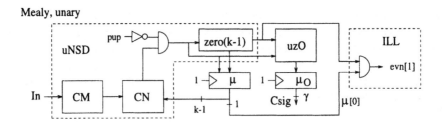

Figure 8.6: Generating event signal *evn*[1], illegal instruction in combination with the four types of control automata

The overflow signal *ovf* of the ALU should only be reported in signed arithmetical operations, i.e., **add, addI, sub,** and **subI**. Therefore, the control automaton generates the control signal *ovf?* during the computation cycle of these instructions. However, on an inactive mask bit $(SR[6] = 0)$, the arithmetic overflow is masked, and must be dropped. Thus, signal *overflow* is produced as

$$evn[6] = overflow = ovf \wedge (ovf? \wedge SR[6]).$$

**Cost and Delay**   Thus, the circuit *Event* which generates the internal event signals has the following cost:

$$C_{Event} = C_{ILL} + 5 \cdot C_{and} + C_{or}.$$

Since the control signals *Csig* and *mis*, the memory status signal *pagefault*, and the overflow signal *ovf* of the ALU environment have the accumulated delays $A_{CON}(Csig)$, $A_{CON}(mis)$, $A_{Menv}(mstat)$ and $A_{ALUenv}(ovf)$ (these times are derived in sections 6.4.2 and 8.1.3), the event signals are valid

$$
\begin{aligned}
A_{Event} = \quad & \max\{A_{ILL}, A_{Menv}(mstat) + D_{and}, \\
& \max\{A_{CON}(Csig) + D_{or}, A_{CON}(mis)\} + D_{and}, \\
& \max\{A_{CON}(Csig) + D_{and}, A_{ALUenv}(ovf)\} + D_{and}\}
\end{aligned}
$$

delays after the start of a cycle.

### Cause Environment

The hardware of this environment is shown in figure 8.7. As mentioned earlier (page 148), we also initiate a reset interrupt on powerup $(pup = 1)$. The input signals *pup* and *reset* are therefore combined by an OR gate.

Internal event signals are only active for one cycle, and that can also be the case for the reset and powerup signal. Therefore, the cause environment accumulates these event signals using one OR gate per event line. Under the control of signal *CAclr*, the corresponding bits of the cause register are cleared at jumps to the ISR. We will have to show, that this does not lead to lost reset or internal events.

The interrupt signals of external I/O devices are active till the interrupt is serviced. Thus, the cause register neither needs to accumulate these bits nor to clear them.

The cost and the cycle time of the cause environment, i.e., circuit GenEvn and register CA, are

$$
\begin{aligned}
C_{CAenv} &= C_{ff}(32) + C_{mux}(7) + 8 \cdot C_{or} + C_{Event} \\
t_{CAenv} &= \max\{A_{Event}, D_{or}, A_{CON}(Csig) + D_{mux}\} + D_{or} + \Delta.
\end{aligned}
$$

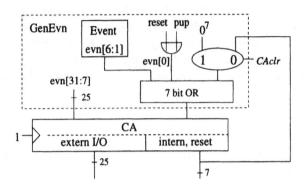

Figure 8.7: Circuit of the cause environment

**Processing of the Caught Event Signals**

We forward the content of the cause register CA and of the status register SR to a circuit CaPro (cause processing) of the control unit which processes the caught event signals. It generates the following signals

$$MCA[i] = \begin{cases} CA[i] \wedge SR[i] & ; \text{if } i \geq 7 \\ CA[i] & ; \text{otherwise} \end{cases}$$

$$int = \bigvee_{i=0}^{31} MCA[i]$$

$$intov = CA[6]$$

$$intm = CA[2] \vee CA[4] \stackrel{\wedge}{=} misa \vee pfls$$

and forwards $MCA[31:0]$ to the special purpose register ECA, and the other signals to the control automaton. Signal $int$ indicates that in the current process an interrupt is pending which is not masked. This signal is used in the fetch state to initiate a jump to ISR. Signal $intov$ indicates that a nonmasked arithmetic overflow has occurred. This signal is tested during write back cycles. Signal $intm$ indicates problems during the current memory access. This signal is tested during the memory access of load and store instructions.

Arranging the OR gates in a balanced binary tree, the cause processing can be done at the following cost and delay

$$C_{CaPro} = 32 \cdot C_{or} + 25 \cdot C_{and}$$

$$A_{CaPro} = D_{and} + 5 \cdot D_{or}.$$

**Extensions of the Control Automaton**

Table 8.6 summarizes the new control signals introduced in the extended data paths. After the modification of the FSD of the DLX machine, we must figure

| Name | Meaning |
|------|---------|
| PCsel | set the input of the PC to SISR |
| spr, spw | read/write signals of the special purpose register file, used on special move instructions |
| Xice | for $X \in \{SR, ESR, EPC, ECA, EMAR\}$, update register X independent of instruction movei2s |
| XDioe | for $X \in \{SR, ESR, EPC\}$, read out register X independent of instruction moves2i |
| SRsl1, SRsl2 | select signals of the two SR multiplexers |
| CAclr | clear the cause register |
| ls | memory cycle during a load or store instruction |
| trap | user requested interrupt |
| ovf? | test of arithmetic overflow is required |

Table 8.6: Additional control signals of the modified DLX hardware

out whether all signals are necessary or whether some of them can be grouped together.

**States of the FSD**  To support interrupt handling, the FSD of the DLX machine requires 10 new states. They are listed in table 8.7. States Movs2i and Movi2s realize the execute cycle of the corresponding instruction. The sole purpose of state TRAP is to generate the event signal $evn[5]$ for the trap interrupt. The sequence of states RFE1, RFE2 realizes in an atomic way the return from an exception.

The sequence of states JISR1, JISR2 realizes in an atomic way the jump to the ISR in the case that the $PC$ has the value at which the program is to be resumed, or in the case that the program is to be aborted. If the PC is already incremented, but the interrupted instruction should be repeated, the sequence of states JISR(-1), JISR1, JISR2 realizes the jump to the ISR. This situation occurs on memory problems during the memory access of load or store instructions (memory problems during fetch are recognized *before incrementing* the PC).

The states Alu0 and AluI0 realize the ALU cycle of the arithmetic instructions add, sub respectively addI, subI. Compared to the states Alu and AluI, they require the additional control signals *ovf?, SRsl1, SRsl2* and *SRice* in order to detect an arithmetic overflow and to catch the overflow flag of the ALU in register SR.

Some old states must generate a few extra control signals. That is also captured in table 8.7. The states Load and Store indicate the data access to main memory by the new signal *ls*. During instruction fetch, signal *fetch* indicates the memory access, but this signal was also generated in the original design.

| State | Meaning | Active Control Signals |
|-------|---------|------------------------|
| Movs2i | $C = SA$ | spr, Cce |
| Movi2s | $SA = A$ | AS1doe, 0S2doe, add, ALUzrDdoe, spw |
| Trap | $evn[5] = trap$ | trap |
| RFE1 | $SR = ESR$ | ESRDioe, SRice |
| RFE2 | $PC = EPC$ | EPCDioe, PCce |
| JISR(-1) | $PC = PC - 4$ | PCS1doe, 4S2doe, add, test, ALUzrDdoe, PCce |
| JISR1 | $EPC = PC - 0$ | PCS1doe, 0S2doe, add, ALUzrDdoe, EPCice |
| JISR2 | $ESR = SR,$ $PC = SISR,$ $EMAR = MAR,$ $ECA = MCA,$ $SR = 0$, clear $CA$ | SRDioe, ESRice, PCsel, PCce, EMARice, ECAice, CAclr, SRice, SRsl2 |
| AluO | $C = A\ op\ B$ test overflow | *like Alu* ovf?, SRsl1, SRsl2, SRice |
| AluIO | $C = A\ op\ co$ test overflow | *like AluI* ovf?, SRsl1, SRsl2, SRice |
| Extended signal lists | | |
| Load Store | memory cycle | *in addition* ls |

Table 8.7: New and modified states of the FSD. SA is the special purpose registers specified by the instruction word, MCA indicates the masked content of the cause register, i.e., $MCA = CA \wedge SR[31 : 7]1^7$.

| Group Name | Corresponding States | Signals |
|:----------:|:--------------------:|---------|
| JISR2 | JISR2 | ESRice, SRDioe, PCsel, EMARice, ECAice, CAclr |
| ovf? | AluO, AluIO | ovf?, SRsl1 |

Table 8.8: Grouping of the new control signals

**Inputs and Outputs of the Automaton**  Table 8.7 indicates, that 8 of the 17 new control signals can be grouped in two groups, because they are active during exactly the same states. These two groups are listed in table 8.8. The signals *spr*, *spw*, *trap*, *ESRDioe*, *EPCDioe*, *EPCice*, and *JISR2* are only active during one state. Thus, they could directly be taken from the state of the control automaton. However, in our model, all those signals count as outputs of the control automaton and therefore, the number $\gamma$ of standard control signals runs at $28 + (17 - 8 + 2) = 28 + 11 = 39$.

The functionality of the control unit also depends on the interrupt events. The preprocessed signals *int*, *intov* and *intm* are new input signals of the control automaton. The clock signal of the PC is now provided directly by the control automaton. Signal *PCsce* is therefore replaced by signal *PCce*. Furthermore, we replaced the memory status signal *busy* by signal *nready*. So, the control unit now has $\sigma = 14 + 3 = 17$ input signals.

**FSD and Disjunctive Normal Forms**  The arcs of the new FSD (figure 8.8) are still labeled with disjunctive normal forms. Most of the monomials remain the same, *busy* and */busy* got replaced. Since the FSD now differentiates four types of arithmetic/logical operations instead of two, the monomials $D2$ and $D5$ must be changed. Table 8.9 lists the modified as well as the additional nontrivial disjunctive normal forms. Once again, the DNFs also make use of the extension of the DLX instruction set, introduced in table 6.15.

In our original DLX design, the longest monomial comprises 10 literals, it now comprises $l_{max} = 12$. After the modification of the DLX design, the set $M$ contains $\#M = 36$ nontrivial monomials instead of 22. Summing up the length of all these monomials yields

$$l_{sum} := \sum_{m \in M} l(m) = 174.$$

The table 8.9 also lists the weight of all arcs with nontrivial DNF, the other arcs have weight 1. Table 8.10 lists the fanin of all states $z$ of the FSD, i.e., the accumulated weight of all arcs ending in node $z$. State WBR has the highest fanin of all states different from fetch, namely

$$\text{fanin}_{max} := \max\{\text{fanin}(z) \mid z \neq z_0\} = 6.$$

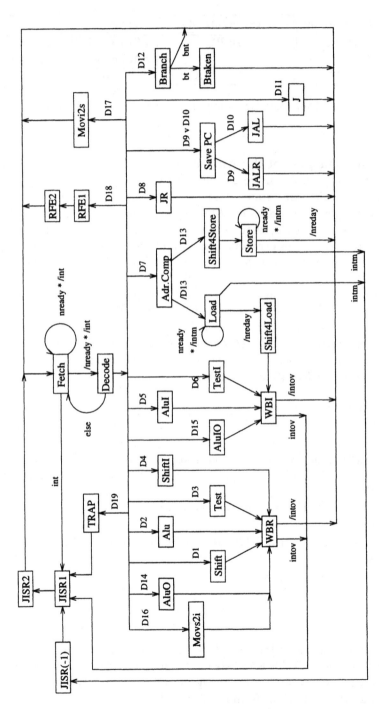

Figure 8.8: FSD of the DLX design supporting interrupts ($k = 32$ states)

| Nontrivial DNF | Target State | Monomial $m \in M$ | | Length $l(m)$ | Weight of arc |
|---|---|---|---|---|---|
| | | IR[31..26] | IR[5..0] | | |
| Unchanged DNFs | | | | | |
| D1 (*) | Shift | 000000 | 0001** | 10 | 1 |
| D3 (*) | Test | 000000 | 101*** | 9 | 1 |
| D4 (*) | ShiftI | 000000 | 0000** | 10 | 1 |
| D6 (*) | TestI | 011*** | ****** | 3 | 1 |
| D7 | Adr.Comp | 100*0* | ****** | 4 | 3 |
| | | 10*0*1 | ****** | 4 | |
| | | 10*00* | ****** | 4 | |
| D8 | JR | 010110 | ****** | 6 | 1 |
| D9 | JALR | 010111 | ****** | 6 | 1 |
| D10 | JAL | 000011 | ****** | 6 | 1 |
| D9 ∨ D10 | Save PC | like P11 and P12 | | | 2 |
| D11 | J | 000010 | ****** | 6 | 1 |
| D12 | Branch | 00010* | ****** | 5 | 1 |
| D13 | Shift4Store | **1*** | ****** | 1 | 1 |
| /D13 | Load | **0*** | ****** | 1 | 1 |
| bt | Btaken | AEQZ ·/IR[26] | | 2 | 2 |
| | | /AEQZ ·IR[26] | | 2 | |
| bnt | Fetch | /AEQZ ·/IR[26] | | 2 | 2 |
| | | AEQZ ·IR[26] | | 2 | |
| Modified or new DNFs | | | | | |
| D2 (*) | Alu | 000000 | 1001** | 10 | 2 |
| | | 000000 | 100**1 | 10 | |
| D14 | AluO | 000000 | 1000*0 | 11 | 1 |
| D5 (*) | AluI | 0011** | ****** | 4 | 2 |
| | | 001**1 | ****** | 4 | |
| D15 | AluIO | 0010*0 | ****** | 5 | 1 |
| D16 | Movs2i | 000000 | 010000 | 12 | 1 |
| D17 | Movi2s | 000000 | 010001 | 12 | 1 |
| D18 | RFE | 111110 | ****** | 6 | 1 |
| D19 | TRAP | 111111 | ****** | 6 | 1 |
| int, intm, intov, /intov, /nready | | | | 1 | 1 |
| nready * /int, /nready * /int, nready * /intm | | | | 2 | 1 |
| Accumulated length of $m \in M$: $\sum_{m \in M} l(m)$ | | | | 174 | |

Table 8.9: Nontrivial disjunctive normal forms (DNF) of the new FSD. DNFs marked with a star make use of the extended instruction set.

| Fanin | States of FSD |
|---|---|
| 15 | Fetch $(= z_0)$ |
| 6 | WBR |
| 5 | JISR1 |
| 4 | WBI |
| 3 | Adr.Comp |
| 2 | Alu, AluI, SavePC, Load, Store, Btaken, JISR(-1) |
| 1 | remaining 20 states |
| 52 | $S_{fanin}$, accumulated fanin of states $\neq$ fetch |

Table 8.10: Fanin of all states in the FSD of the DLX

| MR | 1 | MW | 1 | MARce | 1 | MDRce | 2 |
|---|---|---|---|---|---|---|---|
| MARS1doe | 2 | MDRS1doe | 1 | IRce | 1 | PCce | 9 |
| ABce | 1 | Cce | 11 | IRS2doe | 8 | 0S2doe | 5 |
| PCS1doe | 8 | 4S2doe | 2 | ALUzrDdoe | 15 | ALUzcDdoe | 2 |
| AS1doe | 12 | BS1doe | 1 | BS2doe | 4 | SHDdoe | 4 |
| RFw | 4 | shift | 1 | fetch | 1 | Jlink | 2 |
| Jjump | 2 | Itype | 4 | test | 3 | add | 11 |
| ovf? | 2 | ls | 2 | spw | 1 | spr | 1 |
| trap | 1 | ESRDioe | 1 | EPCDioe | 1 | EPCice | 1 |
| SRice | 4 | SRsl2 | 3 | JISR2 | 1 | | |

Table 8.11: Frequency of the active control signals in the new FSD

**Frequency of the Control Signals** is another relevant parameter of the control automaton. For each control signal $O_i$, we have to determine the number $\nu_i$ of states in the FSD in which it is active. Table 6.17 lists these frequencies for the original FSD. The new FSD only has a few new states. Their active control signals are listed in table 8.7. Adding boths information yields the frequencies of the active control signals in the new FSD (table 8.11).

Signal $ALUzrDdoe$ still has the highest frequency, which now runs at $\nu_{max} = 15$. Summing up the frequencies $\nu_i$ of all the 39 control signals yields

$$\nu_{sum} := \sum_{1 \leq i \leq 39} \nu_i = 137.$$

**Cost and Delay** Using the Mealy and Moore implementations of chapter 4, only minimal information is required to estimate cost and delay of the control automaton. In the previous paragraphs, we derived all the relevant parameters. Table 8.12 lists all of them and compares their new values with the values of the original DLX design.

| Symbol | Meaning | Old Value | New Value |
|---|---|---|---|
| $\sigma$ | # inputs of CM | 14 | 17 |
| $\gamma$ | # output signals of O | 28 | 39 |
| $k$ | # states of FSD | 22 | 32 |
| $\zeta$ | $\zeta = \lceil k \rceil$ | 5 | 5 |
| Frequency of the active control signals in the FSD | | | |
| $\nu_{max}$ | maximal | 10 | 15 |
| $\nu_{sum}$ | accumulated | 93 | 137 |
| nontrivial monomials $m \in M$ | | | |
| $\#M$ | number | 22 | 36 |
| $l_{max}$ | maximal length of $m$ | 10 | 12 |
| $l_{sum}$ | accumulated length | 97 | 174 |
| Fanin of nodes ($\neq$ fetch) in FSD | | | |
| $\text{fanin}_{max}$ | maximal | 4 | 6 |
| $\text{fanin}_{sum}$ | accumulated | 32 | 52 |

Table 8.12: Parameters of the DLX control automaton. The old values correspond to the DLX without interrupts and the new values to the DLX with interrupts.

To estimate cost and delay of the control automaton, we only have to substitute the values of the parameters in the formulae derived in sections 4.2 (Moore implementations) and 4.3 (Mealy implementations).

## 8.2 Cost and Cycle Time

In the previous sections, we derived all the hardware modifications necessary to support interrupts and also developed formulae to estimate the cost and the delay of the new and of the modified environments. Appendix B.8 lists the corresponding C routines. The remaining cost and run time formulae (respectively C routines) can directly be taken from the original design. We now analyze the impact of the hardware extension on the cost and the cycle time of the DLX.

### 8.2.1 Hardware Cost

#### Cost of the Data Paths

Due to the modifications, the DLX data paths got one additional environment and now comprise the following seven environments: arithmetic logical unit, shifter, instruction register, general purpose registers, program counter, main memory, and *special purpose registers*. The cost of the 32-bit wide data paths

|          || ALU  | SH   | IR  | GPR   | PC  | M    | SPR  || DP    |
|----------||------|------|-----|-------|-----|------|------||-------|
| Motorola || 1858 | 1338 | 621 | 5406  | 672 | 930  | 2272 || 13097 |
| Venus    || 2182 | 1378 | 813 | 21627 | 864 | 1250 | 3072 || 31186 |

Table 8.13: Cost of the DLX data paths and of all its environments

can therefore be expressed as

$$C_{DP} = C_{ALUenv} + C_{SHenv} + C_{IRenv} + C_{GPR} + C_{PCenv} + C_{Menv} + C_{SPRenv}.$$

Most of the original environments stay the same. There are only minor changes in the environments of the program counter and of main memory. Table 8.13 lists the cost of the data paths and all its environments under Motorola technology [NB93] and Venus technology [HNS86].

Under Motorola (Venus) technology, the new environment of the special purpose registers roughly accounts for 17% (10%) of the cost. With 41% (69%) of the cost of the data paths, the environment of the general purpose registers still holds the lion's share. However, that is 7 to 8% less than in the DLX design without support for interrupt handling.

## Cost of the Control Unit

As indicated in figure 8.5, the core of the control unit is a Moore control automaton. The automaton computes the next state and generates the standard control signals $Csig$. In addition, the control unit comprises the following circuits:

- circuit MC, it generates the write signals of main memory and the status signal *mis*

- circuit SPRCon, it selects the clock and driver enable signals of the environment SPR

- circuit CaPro, it processes the content of the cause register and generates $MCA$, *int*, *intov* and *intm*

- the cause environment with circuit GenEvn which generates and internal event signals.

As in section 6.3.2, let $C_{auto}$ be the cost of the control automaton. The cost of the control unit can then be expressed as

$$C_{CON} = C_{auto} + C_{MC} + C_{SPRCon} + C_{CaPro} + C_{CAenv}.$$

For the control automaton, we provide the four standard implementations, namely Moore/Mealy implementation with unary/binary coded states. Table 8.14 lists the cost $C_{CON}$ of the control and the cost $C_{DLX}$ of the DLX hardware for all four implementations under both technologies.

| Implementation | | Moore | | Mealy | | Register |
|---|---|---|---|---|---|---|
| Coding of the states | | unary | binary | unary | binary | $C_{ff}(\gamma)$ |
| Motorola | $C_{CON}$ | 1691 | 1626 | 2001 | 1990 | 312 |
| technology | $C_{DLX}$ | 14788 | 14723 | 15098 | 15087 | |
| Venus | $C_{CON}$ | 1947 | 1774 | 2413 | 2294 | 468 |
| technology | $C_{DLX}$ | 33133 | 32960 | 33599 | 33480 | |

Table 8.14: Cost of the hardwired DLX control and of the DLX hardware

The Mealy implementation requires an additional $\gamma$-bit register for the standard control signals and a few additional gates for the powerup mechanism. Thus, under Motorola technology, the Mealy implementation of the control is 18-22% more expensive than the Moore implementation. For Venus technology, the cost increase is about 7% higher due to very expensive registers.

The FSD comprises $k = 32$ states. According to section 4.4, at $k = 32$ binary coding of the states should improve the cost of the control under either technology. For the Moore implementations, the cost reduction amounts to 4% (9%). For the Mealy implementation, the reduction is shorter by about 4%.

The figures of table 8.14 further indicate that the control requires between 11-13% (5-7%) of the cost of the 32-bit DLX design under Motorola (Venus) technology. Thus:

**Maxim 8.1 (Cost of Hardwired Control)** *The hardwired control of a non-pipelined DLX machine with interrupt handling roughly accounts for 12% of the total cost of the DLX machine with 32-bit wide data paths and 32 general purpose registers, at least under Motorola technology.*

*Technologies with very expensive RAMs like Venus (a RAM cell costs 12 gate equivalents) cut this fraction by half.*

Changes in the implementation of the control automaton, like switching from unary to binary coded states, or replacing a Moore implementation by a Mealy implementation, have virtually no impact on the total cost of the DLX machine (less than 2.5%).

## 8.2.2 Cycle Time

The formulae to estimate the cycle times of the data paths, of the memory system, and of the control unit can largely be taken from the original DLX design.

### Delay of the Control Signals

In a first step, we evaluate the accumulated delay of the control signals, which the control unit forwards to the data paths via bus Cout. The standard control signals $Csig$ are generated by the control automaton. The memory write

| Approach | Motorola | | | | Venus | | | |
|---|---|---|---|---|---|---|---|---|
| | Moore | | Mealy | | Moore | | Mealy | |
| Coding | un. | bin. | un. | bin. | un. | bin. | un. | bin. |
| $A_{CON}(Csig)$ | 8 | 15 | 0 | 0 | 4 | 8 | 0 | 0 |
| $A_{CON}(Mw)$ | 14 | 21 | 11 | 11 | 7 | 11 | 6 | 6 |
| $A_{SRPCon}$ | 12 | 19 | 9 | 9 | 6 | 10 | 5 | 5 |
| $A_{CaPro}$ | 9 | 9 | 9 | 9 | 6 | 6 | 6 | 6 |

Table 8.15: Delay of the control signals. On 'un.' ('bin.'), the states of the control automaton are coded in unary (binary).

signals $Mw$ come from circuit MC, signals MCA (masked content of register CA) come from circuit CaPro, and finally, the clock and select signals of the special purpose register environment come from circuit SPRCon. These control signals are valid $A_{CON}(Csig)$, $A_{CON}(Mw)$, $A_{CaPro}$, respectively $A_{SPRCon}$ gate delays after the start of the cycle. Table 8.15 lists all these delays under different implementations of the control and under both technologies.

Using a Mealy approach, the *standard* control signals are directly taken from a register in the control unit, and so they are valid immediately after the new cycle started. Switching between binary and unary coding of the states therefore has no impact on the accumulated delay of the control signals $Csig$.

Using a Moore implementation with unary coded states, the standard control signals first have to pass the output circuit O. That adds 8 (4) gate delays. The decoder in the binary version almost doubles this delay.

The control circuits MC and SPRCon have some standard control signals as inputs, but the computation of these circuits can partially be overlapped with the generation of the standard control signals. That is the reason why these signals are not delayed by a constant amount of time. The additional delay can be up to 11 (6) gate delays.

### Cycle Time of the Data Paths

Compared to the original DLX data paths of chapter 6, interrupt handling only adds the environment of the special purpose registers and causes some minor changes in the environments of the PC and of the main memory. Those changes only impact the accumulated delay of the memory status signal *nready* and of the clock signal $PCce$.

As in section 6.4.2, we use a simplified view of the data paths (figure 8.9) in order to spot all the cycles through the data paths. The hardware extension cause three new cycles, namely:

- from a special purpose register though circuit SPRsel back to the registers in the SPR environment

- from inputs MCA to a register in the SPR environment

- from the SPR environment via D bus and circuit Regw into an register.

All the other changes do occur in the control unit, in circuit RegCE which provides the clock signals to the registers Reg, or in circuit Regw which connects the D bus to the data inputs of the registers.

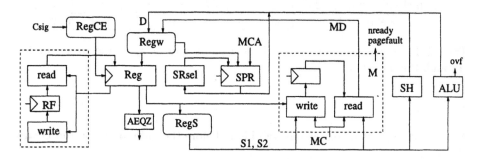

Figure 8.9: Simplified view of the DLX data paths. New paths are in bold type

In section 8.1.3, we derived for the first of the new cycles the time

$$T_{SRsel}(SR; SR) = \max\{A_{CON}(Csig) + 2 \cdot D_{mux}(32), A_{SPRCon}\} + \Delta.$$

The cause processing circuit CaPro of the control provides the data MCA. Thus, the cycle time of the second new cycle runs at

$$T_{SPR}(MCA) = A_{CaPro} + D_{SRP}(MCA) + \Delta.$$

During the third of the new cycles, a special purpose register places its data on the D bus. That takes time $A_{SPR}(; D)$. Circuit Regw then forwards the data to the inputs of the registers. Thus, this cycle has a cycle time of

$$T_{SPR}(SPR; D) = A_{SPR}(; D) + D_{Regw} + \Delta.$$

Circuit RegCE became simpler, because all clock signals of the registers Reg are now standard control signals, i.e., they are generated by the control automaton. Thus,

$$T_{RegCE} = A_{CON}(Csig) + \Delta.$$

$D_{Regw}$ indicates the delay required to forward data from the busses D and DM to the registers. So far, this circuit had delay $D_{mux}(32)$. Due to the design extension, we also have to consider the setup time of the special purpose registers. However, the new setup time also equals the delay of a 32-bit multiplexer. The changes in the PC environment do not impact its setup time and delay. Thus,

$$D_{Regw} = \max\{D_{mux}(32), D_{PCenv}(D; 32), D_{SPR}(D; )\} = D_{mux}(32).$$

| Cycle Time | | GPR | SH | ALU | | SPRenv | | | DP |
| Implementation | | | | ZR | ZC | SRsel | ;D | MCA | |
|---|---|---|---|---|---|---|---|---|---|
| M | Moore, un. | 35 | 46 | 75 | 88 | 17 | 21 | 17 | 88 |
| | Moore, bin. | 42 | 53 | 82 | 95 | 24 | 28 | 17 | 95 |
| | Mealy | 27 | 38 | 67 | 80 | 14 | 18 | 17 | 80 |
| V | Moore, un. | 22 | 37 | 47 | 53 | 13 | 14 | 11 | 53 |
| | Moore, bin. | 26 | 41 | 51 | 57 | 17 | 18 | 11 | 57 |
| | Mealy | 18 | 33 | 43 | 49 | 10 | 13 | 11 | 49 |

Table 8.16: Cycle time of the DLX data paths and delay of the major cycles through the data paths

Note, that all control signals are valid long before the data arrive on the D and DM bus.

$T_{DP}$ indicates the cycle time of the data paths. This time now also has to include the cycle time of the environment SPR. Consequently, the new formula is

$$T_{DP} = \max\{T_{RegCE}, T_{GPR}(; RF), T_{GPR}(RF;), T_{SH}, T_{ALU}(ZR),$$
$$T_{ALU}(ZC), T_{SRsel}(SR; SR), T_{SPR}(MCA), T_{SPR}(SPR; D)\}.$$

Table 8.16 lists the cycle time of the DLX data paths and of all the relevant cycles within the data paths. Almost all cycles first read the standard control signals $Csig$. Thus, the accumulated delay $A_{CON}(Csig)$ adds to their cycle time. Using a Mealy implementation of the control automaton therefore speeds up the cycle time of the data paths in the same way as the computation of the control signals.

The times of the old cycles through the data paths remain the same, and the cycles through the environment of the special purpose registers (new cycles) are far shorter than most of the old ones. Thus, the time critical path still goes through the adder and the comparator. Under Motorola (Venus) technology, the comparator accounts for 18% (13%) of the delay of this critical path. This cycle is still over 39 (27)% slower than any other cycle in the data paths which avoids the carry look-ahead adder.

In analogy to the DLX without interrupt handling, the design modifications of chapter 7 can be used to speed up the arithmetic unit and condition test of the current DLX design. Since these trade-off analyses mainly confirm the results of the previous chapter, we leave the execution of these analyses to the reader. Once again, the design variants with a conditional sum adder prove to have a higher quality than the current DLX design, if there is no fanout restriction.

**Cycle Time of the Control**

After the hardware extension, the control unit has to update two registers, i.e,
the status of the control automaton and the cause register CA of the cause
environment (see figure 8.5).

The next state unit of the control automaton got a few new inputs, namely
the status signal *nreday* of the memory environment and the outputs of the
cause processing circuit CaPro. The input signals now have an accumulated
delay of

$$A(In) = \max\{A_{GPR}(AEQZ), A_{Menv}(nready), A_{CaPro}\}$$

gate delays. The cycle which ends in the cause register CA require time $T_{CA}$, as
derived on page 8.7. Thus, the cycle time of the control unit runs at

$$T_{CON} = \max\{T_{auto}, T_{CAenv}\}.$$

**Cycle Time of the DLX**

The hardware extensions for handling interrupts do not impact the cycle time of
the main memory system, and its cycle time is still modeled by the same formula
as in the original DLX design:

$$T_M = \max\{A_{CON}(Mw), A_{RegS}\} + d_{mem} + D_{Regw} + \Delta.$$

Further more, the hardware modifications only impact indirectly the formula of
the cycle time of the whole DLX machine, which can be expressed as

$$T_{DLX} = \max\{T_{DP}, T_{CON}\}.$$

Memory accesses still require $\lceil T_M/T_{DLX} \rceil$ cycles. The figures of table 8.17
indicate, that memory accesses can be performed in one machine cycle, if the
memory access time $d_{mem}$ stays under 68% of the cycle time of the data paths.

Among others, the cause environment gets the signals $ovf$ and the memory
status signals as inputs. The table 8.17 lists the accumulated delay of these
signals as well. The memory status time $d_{mstat}$ can grow up to 75.5% of the
cycle time of the data paths before $d_{mstat}$ causes the cause environment to be
slower than the data paths.

The memory status time also impacts the cycle time of the control automaton.
However, in combination with a Mealy (Moore) implementation of the control
automaton, the cycle time of the control unit does not dominate the cycle time
of the whole machine as long as the memory status time stays under 40% of
the cycle of the data paths. The DLX with interrupt handling can also tolerate
a larger memory status time, when we adapt the modification of the control
automaton presented in section 6.5 (see exercise 8.1).

These results are similar to the results of the original DLX design, and so we
can generalize rule 6.2 as follows:

| Implementation | | Accumulated Delay $A$ | | Cycle Time $T$ | | | | |
|---|---|---|---|---|---|---|---|---|
| | | ovf | nready | $T_{auto}$ | $T_{CA} =$ max$\{A$ | , $B\}$ | $T_{DP}$ | $T_M$ |
| M | Moore, un. | 64 | 16 | 49 | 21 | 73 | 88 | 21 |
| | Moore, bin. | 71 | 23 | 57 | 28 | 80 | 95 | 28 |
| | Mealy | 56 | 13 | 48 | 18 | 65 | 80 | 18 |
| V | Moore, un. | 37 | 8 | 28 | 13 | 44 | 53 | 14 |
| | Moore, bin. | 41 | 12 | 32 | 17 | 48 | 37 | 18 |
| | Mealy | 33 | 7 | 28 | 12 | 40 | 49 | 13 |
| Displacement | | | | $+d_{mstat}$ | | | | $+d_{mem}$ |

Table 8.17: Cycle time of the control, the data paths and the memory environment. M indicates Motorola technology and V Venus technology.

**Maxim 8.2 (Cycle Time of Hardwired Control)** *If the hardwired control of the non-pipelined DLX (with or without interrupt handling) is realized as Mealy implementation of a Moore automaton, and if the memory status time $d_{mstat}$ is small, the control does not lie on the time critical path of the design, even when the automaton is implemented without any optimization.*

## 8.2.3 Design Evaluation

Lets assume that the memory status time stays under 40% of $D_{DP}$. Then, under Motorola (Venus) technology, the DLX machine with interrupt handling can run at the same cycle time as before: $t_C \geq 80$ (49). A Moore implementation of the control automaton still slows the DLX down by 4-15 delays, that is 10 to 20% of the cycle time. On the other hand, a Moore implementation of the control automaton reduces the cost of the DLX by less than 2.5%. Consequently, the DLX with a Mealy implementation has at least a 7.5% better price/performance ration independent of the benchmark application.

When comparing the quality $Q_{q,A}$ of the DLX designs $A$ under different implementations of the control automaton (figure 8.10), the Mealy variant again proves to have highest quality, at least for quality parameters $q$ less than 0.82. Thus, we can generalize rule 6.3 as follows:

**Maxim 8.3** *Under a realistic quality metric $Q_q$ with $q \in [0.2 : 0.5]$, the design variant of the non-pipelined DLX which uses a Mealy implementation of the control automaton has a higher quality than the design variants which use a Moore implementation of the control automaton.*

Table 8.18 compares the cost and cycle time of the DLX designs with and without interrupt handling. Thus, support for interrupts roughly doubles the cost of the control, but it has a less severe impact on the data paths. Under Motorola technology, it increases the cost of the data paths by 21%, and the

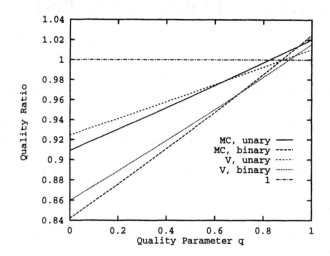

Figure 8.10: Quality ratio of the DLX designs with Moore control automaton relative to the DLX design with Mealy control automaton and binary coded states. MC indicates Motorola technology and V Venus technology

whole DLX hardware becomes 28% more expensive. Under Venus technology with its expensive register file, the support for interrupts only accounts for half the cost increase than under Motorola technology.

**Maxim 8.4 (Supporting Interrupts)** *Adding hardware support for interrupt handling makes the non-pipelined DLX design at most 28% more expensive. The additional functionality does not increase the cycle time of the design, at least if the memory status time is reasonably small.*

| | | Cost | | | Cycle Time |
|---|---|---|---|---|---|
| | | DP | CON | DLX | |
| | Old Value | 10823 | 1014 | 11837 | 80 |
| Motorola | New Value | 13097 | 1990 | 15087 | 80 |
| | Increase | 21% | 96% | 28% | 0% |
| | Old Value | 28112 | 1214 | 29326 | 49 |
| Venus | New Value | 31186 | 2294 | 33480 | 49 |
| | Increase | 11% | 89% | 14% | 0% |

Table 8.18: Comparison of cost and cycle time of the two DLX designs. The old values correspond to the DLX without interrupt handling

## 8.3  Correctness of Interrupt Handling

We will (carefully) formalize and prove that the interrupt mechanism of the DLX architecture outlined in the previous sections handles interrupts correctly. To do so, we have to argue on a cycle by cycle basis, and therefore we number all cycles of a computation by numbers $t = 1, 2, \ldots$. The correctness proof hinges on the following definitions, some of which are very carefully worded.

Note that, masking an interrupt has a fundamentally different effect for internal interrupts (in our case only $ovf$) and external interrupts. If an internal interrupt occurs while it is masked, it is ignored altogether. If an I/O interrupt is masked, it is temporarily ignored and should receive service once it is unmasked.

**Definitions:**

1. *Interrupt $j$ occurs* in cycle $t$, if signal $evn[j]$ switches from 0 to 1 during cycle $t$.

2. Interrupt $j$ is *caught* in cycle $t$ if bit CA[j] switches from 0 to 1 at the end of cycle $t$. This is the case exactly if interrupt $j$ occurs in cycle $t$ and $j$ is not a masked internal interrupt.

3. We denote by $eca(t)$ the value of ECA stored in the top frame of the interrupt stack in cycle $t$.

4. We define inductively the *interrupt level $il(t)$* in cycle $t$:

   - If at the beginning of cycle $t$ the interrupt stack is empty and the control automaton is in state fetch, then $il(t) = 32$. This is the interrupt level of the uninterrupted program.
   - If in cycle $t$ the machine is in state JISR2 then
     $$il(t + 1) = \min\{i \mid ECA(t + 1)[i] = 1\},$$
     where ECA(t+1) indicates the value of register ECA during cycle $t+1$. In this case, interrupt service routine *ISR(il(t))* gets interrupted by ISR(il(t+1)).
   - If in cycle $t$ the machine is in state REF2, and in cycle $t + 1$ the interrupt stack is not empty, then
     $$il(t + 1) = \min\{i \mid eca(t + 1)[i] = 1\} \quad .$$
     In this case interrupt service routine *ISR(il(t))* terminates.
   - In all other cases $il(t + 1) = il(t)$.

5. We say interrupt $j$ *receives service* in cycle $t$ if $il(t) = j$ or if $il(t) = j'$, where interrupt $j'$ aborts the program.

6. Interrupt $j$ is *pending* in cycle $t$ if it was caught in some cycle $t' \leq t$ and has not received service in cycles $\{t', \ldots, t\}$.

7. We define inductively for each cycle $t$ the *instruction $I(t)$ executed in cycle $t$*: A new instruction starts in cycle $t$, if in cycle $t - 1$ the machine is not in state `Fetch` and in cycle $t$ the machine is in state `Fetch`. In all other cases $I(t) = I(t - 1)$. This defines an ordering on the instructions executed. Because interrupt levels can only change at the same time when new instructions start, this also defines for each instruction $I$ the *interrupt level of instruction I*.

8. For instructions $I$, we denote by $I^+$ the instruction which follows $I$ in the absence of interrupts.

9. The CPU status consists of the following 33 registers: the general purpose registers R1 to R31, the program counter PC, and the status registers SR.

   Note that the ISR saves the cause register CA and the memory address register MAR in addition to the CPU status, but it does not restore these additional registers before it jumps back to the program.

10. We call a set of interrupt service routines *admissible* if the following properties hold:

    (a) Every interrupt service routine terminates, if it is not interrupted.

    (b) The interrupt stack is held on a permanent page of memory. (We will show, that the interrupt stack is of bounded size.)

    (c) The code segments *Save Status* and *Restore Status* of the interrupt service routine ISR are held on a permanent page of memory.

    (d) For $i \leq 5$, the code of interrupt service routines $ISR(i)$ is held on permanent pages of memory.

    (e) For $i \leq 5$, the code of interrupt service routines $ISR(i)$ contains no illegal instructions and does not produce page faults or misaligned memory accesses.

The correctness of the interrupt mechanism outlined in sections 8.1 is (hopefully) captured in

**Theorem 8.5** *We assume, that the interrupt service routines are admissible, that finitely many interrupts occur, and that for all $j \in \{0, \ldots, 31\}$, the interrupt service routine $ISR(i)$ has a finite run time. Then, for all $t$ and all $j \in \{0, \ldots, 31\}$ the following holds:*

1. *Response Time:*

   (a) *If a reset interrupt occurs in cycle $t$, it receives service in cycle $t + WS + 9$ or earlier.[2]*

   (b) *If a maskable external interrupt $j$ occurs in cycle $t$ and it is not masked at the end of instruction $I(t)$, then some interrupt receives service in cycle $t + WS + 9$ or earlier.*

---

[2]Recall that $WS$ is the number of wait states during a memory access.

> (c) Every unmasked internal interrupt that occurs during an instruction is serviced in the next instruction (possibly by the reset routine $ISR(0)$).

2. Priority: Once interrupt $j$ received service, it can only be interrupted by an unmasked interrupt of higher priority or by reset.

3. Finite Space: The interrupt stack is of bounded size.

4. Preciseness: Suppose, interrupt $j$ receives service for the first time in cycle $t + 1$ and instruction $I$ equals $I(t)$.

> (a) If interrupt $j$ is an I/O interrupt, instruction $I$ does not change the CPU status and the program is resumed at instruction $I$.

> (b) If interrupt $j$ is an internal interrupt of type repeat instruction $I$ does not change the CPU status and the program is resumed at instruction $I$.

> (c) If interrupt $j$ is an internal interrupt of type continue instruction $I$ updates the CPU status and the program is resumed at instruction $I^+$.

5. Termination: If interrupt $j$ received service, $ISR(j)$ terminates or the program is aborted.

6. Completeness: Every interrupt which is unmasked long enough receives service.

**Proof:** 1.a) By construction of the cause environment, reset interrupts that occur are caught in every cycle. This includes the cycle when the cause register is cleared, because the OR gate is behind the multiplexer. Once bit CA[0] equals 1, it is not cleared until state JISR2 is reached. One cycle later interrupt 0 receives service.

Inspection of the state diagram shows, that at most $WS+9$ cycles after a reset interrupt occurs it receives service. This delay occurs, if the interrupt is caught in the last fetch cycle of a load instruction. The control automaton then takes the following paths before it starts ISR(0): Decode, Adr.Comp, Load ($WS + 1$ cycles), Shift4Load, WBI, Fetch, JISR1, JISR2, Fetch.

1.b) By the interrupt convention, once an I/O-interrupt has occurred its event line stays on until the interrupt has received service. If interrupt j occurs while it is masked, or if it occurs during the last fetch cycle or a later cycle of a movei2s instruction which masks interrupt $j$, then the interrupt is ignored until it is unmasked again. Otherwise, an interrupt receives service after at most $WS + 9$ cycles in analogy to 1.a). Even if interrupt $j$ stays unmasked, higher level interrupts might be caught and receive service instead of interrupt $j$.

1.c) Inspection of the state diagram shows, that several internal interrupts can occur in a single instruction $I$. Table 8.19 lists the possible combinations, and the types of the interrupts.

| Events | Combinations | | | | Type | |
|---|---|---|---|---|---|---|
| Reset | x | x | x | x | external | abort |
| Illegal Instruction | x | | | | internal | abort |
| Misaligned memory access | x | x | | | internal | abort |
| Page fault on fetch | x | | | | internal | repeat |
| Page fault on load/store | | x | | | internal | repeat |
| Trap | | | x | | internal | continue |
| Overflow | | | | x | internal | continue |
| I/O events | x | x | x | x | external | continue |

Table 8.19: Possible combinations of interrupt events which can become active
during the same instruction

Only memory access problems can cause more than one internal interrupt
for a single instruction. In this case, the highest pending interrupt is aborting.
By our definition, if the highest pending interrupt receives service, all the other
pending internal interrupts receive service as well. The design of the memory
system (sections 6.2.6, 8.1.3) and the state diagram guarantee, that the control
automaton jumps to state JISR1.

The internal interrupts 5 and 6 can only occur together with external interrupts. A trap ($j = 5$) is not maskable, but an overflow is. If an unmasked
overflow interrupt ($j = 6$) occurs during instruction $I$, then $I$ is not a *movi2s*
instruction. Thus, interrupt 6 stays unmasked after the instruction. In each
case, the interrupt level of the next instruction is either $j$ or 0 (in case a reset
interrupt occurred). Thus, interrupt $j$ receives service.

2. For maskable interrupts $j$, this follows directly from the construction of the
interrupt service mechanism in section 8.1. During the service of a nonmaskable
interrupt ($j \leq 5$) *no* interrupts other than reset can occur, because the interrupt
service routines are admissible.

3. By part 2 of the theorem, the interrupt stack can have at most 32 frames,
i.e., it needs at most $32 \cdot 35 \cdot 4 = 4480$ bytes.

4.a) The control automaton tests for external interrupts only in state Fetch.
If during any fetch cycle of instruction $I$ an unmasked I/O-interrupt is pending,
the automaton branches to state JISR1. Instruction $I$ does not update the CPU
status and the PC which belongs to $I$ is pushed on the interrupt stack.

4.b) and 4.c) For the internal interrupts 3 to 6 this follows immediately from
the state diagram. Note that for the interrupts pff (page fault on fetch) and pfls
(page fault on load/store) of type repeat, the PC which was already increased
in state decode is decreased again in state JISR(-1).

5. This follows by an easy induction on $j$ from the finite run time of the single
interrupt service routines and part 4.

6. Because we assume that only finitely many interrupts occur, it suffices
to show, that no interrupt which is caught, is lost before it receives service.

For maskable external interrupts this follows directly from the interrupt convention. For reset and internal interrupts it follows from parts 1.a) and 1.c) of the theorem. □

**Theorem 8.6 (Powerup)** *Let the interrupt service routine ISR(0) initialize all registers and RAMs of the data paths. Under the premises of theorem 8.5, the current DLX design starts ISR(0) a few cycles after powerup. When the machine then leaves ISR(0), it is in a well defined status. Except from the cause register CA, this status is always the same.*

**Proof:** On powerup, the external signal *pup* is active for one cycle. Thus:

- Since the cause environment (figure 8.7) combines the external signals *pup* and *reset* via an OR gate to event signal *evn*[0], this event signal is now active and is caught at the end of the cycle.

- The control automata of chapter 4 are designed such that on *pup* = 1 they pull its next state to the initial state $z_0$, which in the current DLX design equals state *fetch*. The Mealy implementation further set the next value of their output register to output signals of state $z_0$. Thus, in the next cycle, the DLX will perform an instruction fetch.

During the second cycle, interrupt 0 is pending and therefore interrupt signal *int* is active. The automaton is in state **Fetch**, and the data paths fetch an instruction at a random address. Since *int* = 1, the control jumps to state *JISR*1. According to theorem 8.5, the DLX machine can now only be interrupted by *reset*, but signal *pup* is now inactive. Interrupt 0 will therefore receive service within a few cycles[3]. Since ISR(0) has a finite run time, the DLX machine finally returns from the interrupt service routine. The control is then in state **Fetch**. Since ISR(0) initializes all registers and RAMs of the data paths, the machine will always be in a well defined status. The cause register may already have caught new interrupts. However, after returning from that initial interrupt, all the other register will have the same value. □

## 8.4 Exercises

8.1 In exercise 6.4 of section 6.5, we described a mechanism which allows the DLX architecture without interrupt support to tolerate a fairly high memory status time.

    1. Adapt this mechanism for the current DLX design. Originally, only external signal *pup* could override the memory status signal *nready*. Depending on the state of the FSD, interrupt signals *int* and *intm* can now also override *nready*.

    2. How does this mechanism impact circuit ILL?

---

[3]This only holds, if the reset button is not pressed constantly

3. What is the maximal memory status time, which the modified DLX design with interrupt support can tolerate?

8.2 The ALU still lies on the time critical path. Thus a conditional sum adder or a modified implementations of the condition test (chapter 7) could improve the cycle time.

1. Analyze the cost/time trade-offs among these six design variants. Assume, that there is no fanout restriction.

2. Repeat the analyses for a fanout restriction of $f = 2$.

3. Compare the results with those of chapter 7.

# Chapter 9

# Microprogrammed Control

So far, we only used hardwired Mealy and Moore automata to control the DLX machines. Microprogrammed (microcoded) control is another common approach which goes back to an idea of Maurice Wilkes [Wil51, WS53] .

Most parts of bit-parallel designs are simple or regular, only control is more complex and difficult. That makes it hard to get the control working right. Wilkes therefore tried to make the control more regular. He developed the concept of *microprogrammed control* which turns the control into a computer itself. To avoid confusion, we use the prefix "macro" to refer to higher level components.

## 9.1 Basic Principles

In the previous chapters, we saw that the control can be modeled by a Moore control automaton with states $z \in Z$, inputs $I = I_1, \ldots, I_\sigma$ and outputs $O = O_1, \ldots, O_\gamma$. For any pair of state $z$ and inputs $I$, its total transition function defines the next state as $\delta(z, I)$. Its output function $\tau$ only depends on the current state but not on the inputs.

For any control automaton, the underlying FSD partitions the execution of a machine instruction in a sequence of basic actions, each of which can be performed in a single cycle. These actions are called *microinstructions* and the sequence of microinstructions used to perform machine instruction $J$ is called the *microprogram* of $J$. The microinstructions correspond one to one to the states of the FSD.

In a microprogrammed (microcoded) control, the microprograms of all machine instructions are stored in the *microprogram ROM* ($\mu$-ROM). This ROM is addressed via a pointer, the *microinstruction PC* ($\mu$-PC). The *microinstruction register* ($\mu$-IR) holds the microinstruction currently executed. In our model, any microinstruction comprises the following three fields (see figure 9.1):

- The $\gamma$ bits $\mu sig$ govern the macro-machine. For any state $z \in Z$, they code the output value $\tau(z)$ of the control automaton.

- The field $\mu adr$ is $\zeta$ bits wide and specifies an address of the microprogram ROM. These bits are used as target address of a microinstruction branch. For a control automaton with $k$ states, $\zeta$ equals $\lceil \log k \rceil$.

- The field $\mu cond$ is $\beta$ bits wide, and is often called *microcondition* or *sequencing data* of the microinstruction. Together with the $\sigma$ inputs of the control automaton, with the address $\mu adr$ and with the value of the current $\mu$-PC, these bits select the next value of the $\mu$-PC.

Thus, the microinstructions are $\alpha = \beta + \gamma + \zeta$ bits wide. The microinstruction word provides one bit per output signal $O_i$ of the control automaton. This coding of the control signals is called *unary coding*.

During the execution of a microinstruction, the microcoded control computes the address of the next microinstruction. It also generates the control signals of the macro-machine and the data paths then perform the corresponding actions. When overlapping the execution of the current and the fetch of the next microinstruction, the microcoded control can schedule one instruction per cycle; the $\mu$-ROM and the data paths of the macro-machine are traversed in parallel.

Like the hardwired control automata of chapter 4, the microcoded control should on powerup ($pup = 1$) also switch to its initial state, i.e., in the next cycle, the microcoded control must execute the microinstruction which corresponds to the initial state $z_0$ of the underlying FSD. We store this microinstruction on address 0.

Figure 9.1 depicts the basic scheme of a microcoded control implemented along these lines. On powerup, it pulls the outputs of the address selection circuit down to zero. This value is the stored into the $\mu$-PC and is used to load the next microinstruction into register $\mu$-IR. Thus, in the next cycle, the microcoded control executes the microinstruction of the initial state $z_0$.

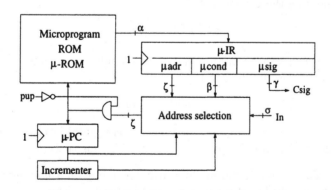

Figure 9.1: Basic scheme of microcoded control

Let $C_{Select}$ denote the cost of the address selection unit; the microcoded control with a $k$-state FSD and with an $\alpha$-bit microinstruction word then costs

$$C_{micro} = C_{ROM}(k, \alpha) + C_{ff}(\alpha) + C_{ff}(\zeta) + C_{Inc}(\zeta) + \zeta \cdot C_{and} + C_{inv} + C_{Select}$$

gate equivalents. Let the inputs $I_1, \ldots, I_\sigma$ be valid $A(In)$ delays after the start of the cycle, and let $D_{Select}$ denote the delay of the address selection unit. Since any external input, including signal *pup* has an accumulated delay of zero, and since an incrementer has at least delay $D_{inv}$, the microcoded control has the cycle time

$$
\begin{aligned}
T_{micro} &= \max\{D_{inv}, \max\{D_{Inc}(\zeta), A(In), A_{CON}(Csig)\} + D_{Select}\} \\
&\quad + D_{and} + D_{ROM}(k, \alpha) + \Delta \\
&= \max\{D_{Inc}(\zeta), A(In), A_{CON}(Csig)\} + D_{Select} \\
&\quad + D_{and} + D_{ROM}(k, \alpha) + \Delta.
\end{aligned}
$$

The outputs and the microcondition of the control are valid $A_{CON}(Csig)$ delays after the start of the cycle. In the current implementation, that equals zero:

$$
A_{CON}(Csig) = 0.
$$

## 9.1.1 The Address Selection Unit

FSDs of control automata usually manage with a few conditional branches. Most of their states have one direct successor, and this transition is taken independent of the inputs of the automaton. A microcoded control usually exploits this property of the FSDs by storing the instructions of a microprogram consecutively. Thus, an incrementer can often compute the address of the next instruction. Although conditional branches are an exception, they are nevertheless necessary on the microinstruction level.

The following basic methods are commonly used in order to compute the *target address*, i.e., the address of the next microinstruction.

1. Incrementing the current $\mu$-PC:

$$
\mu\text{-PC} = \mu\text{-PC} + 1 .
$$

2. Performing an unconditional branch to the address specified in field $\mu adr$ of the microinstruction word:

$$
\mu\text{-PC} = \mu adr .
$$

The microcoded control uses this mechanism (goto) in order to reach microcodes which are shared by several machine instructions.

3. Performing a conditional microbranch depending on a *microbranch condition* $\mu bc$ specified by field $\mu cond$ of the instruction word:

$$
\mu\text{-PC} = \begin{cases} \mu adr & \text{if } \mu bc = 1 \\ \mu\text{-PC} + 1 & \text{otherwise} . \end{cases}
$$

If one uses branch conditions, which are identically false respectively identically true, then the first two mechanisms are obviously special cases of the third mechanism.

4. Performing a single microinstruction loop depending on the microbranch condition:

$$\mu\text{-PC} = \begin{cases} \mu\text{-PC} & \text{if } \mu bc = 1 \\ \mu adr & \text{otherwise} \end{cases}.$$

This is not a special case of the first mechanism, because one can jump from the loop immediately to any place in the microprogram and not just to the next microinstruction.

5. Performing a table lookup. The target address is looked up in a ROM, which is addressed by some of the inputs of the FSD. This mechanism is necessary in order to leave a microinstruction which is shared by several machine instructions. A typical example is the decode state.

These mechanisms suffice to implement the FSD without interrupts from figure 6.18 such that each state corresponds to one microinstruction. If we want to handle interrupts, this cannot possibly work, because we have to perform 3-way branches. It is of course possible, to simulate a 3-way branch by two 2-way branches [HP90]. But this would amount to continuous polling on the microinstruction level. The whole point of interrupts is of course to avoid continuous polling.

Therefore, we permit in field $\mu cond$ the specification of a *second* branch condition $ibc$, called *interrupt branch condition*. Branches due to these conditions will have priority over other branches.

The resulting structure of the address selection unit is shown in figure 9.2. The field $\mu cond$ of the microinstruction word now has three subfields, namely

1. subfield $ibc$ for the specification of interrupt branch conditions,

2. subfield $\mu bc$ for the specification of microbranch conditions, and

3. subfield $\mu type$ which specifies the type of address calculation to be performed (branch, loop or table lookup).

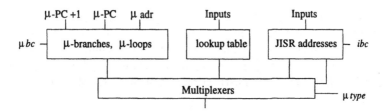

Figure 9.2: Address selection of a microcoded control

Let the microcondition $\mu cond$ code $n_{inter}$ interrupt conditions and $n_{branch}$ conditions of microbranches and microloops, and select $n_{type}$ types of address calculation. In a first approach, the microinstruction word is coded in unary. Then, the field $\mu cond$ is $\beta$ bits wide, with

$$\beta = n_{branch} + n_{inter} + n_{type}.$$

## 9.1.2 Implementation of a Table Lookup

As usual, let $k$ denote the number of states of the FSD, let $\zeta = \lceil \log k \rceil$, and let $I_1, \ldots I_\sigma$ be the $\sigma$ input signals $In$ of the FSD. If we implement FSDs with one microinstruction per state, the microprogram memory has $k$ locations and $\zeta$ address bits.

For any state $z$ of the FSD the next state is a switching function

$$next_z : \{0,1\}^\sigma \to \{0,1\}^\zeta$$

of the inputs of the FSD and can trivially be computed by a $2^\sigma \times \zeta$ ROM. Usually, the functions $next_z$ do not depend properly on all inputs $I_j$ of the FSD. Hence, if function $next_z$ properly depends on $s(z) < \sigma$ inputs, the lookup table for state $z$ can be realized with a $2^{s(z)} \times \zeta$ ROM.

## 9.1.3 Modifications of the Microcoded Control

The modifications of the microcoded control presented in this section aim at reducing the width of the microinstructions and of the $\mu$-ROM.

### Encoding Control Signals

So far, the microcoded control codes the control signals of the macro-machine in unary, i.e., the microinstruction word provides one bit per control signal $O_i$. However, there are some groups of signals with at most one active signal per cycle. Output enable signals of tristate drivers connected to the same bus are the best example. Some busses even require exact one of their enable signals to be active. Such a group of control signals can be coded in binary without loosing any information.

Certainly, the sequencing information $\mu cond$ of the microinstruction can also be encoded. Let the encoded sequencing information require $b$ instead of $\beta$ bits, and let the encoded control signals require $g$ instead of $\gamma$ bits. Thus, encoding reduces the width of the microinstruction and of the $\mu$-ROM from $\alpha = \zeta + \beta + \gamma$ to $a = \zeta + b + g$.

However, the encoded fields must be decoded later on. The decoders can be inserted in front or behind the microinstruction register (figure 9.3).

Let $C_{Select}$, $D_{Select}$, $C_{\mu Dec}$ and $D_{\mu Dec}$ denote cost and delay of the address selection unit and of the decode circuit required to decode the microinstruction. The accumulated delay $A_{CON}(Csig)$ of the control signals usually contributes to the delay $A(In)$ of the inputs of the control automaton; $A(In) \geq A_{CON}(Csig)$.

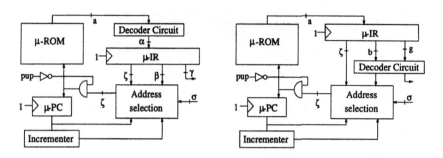

Figure 9.3: Microcoded control with encoded microinstructions

Under these premises, the following formulae express cost and cycle time of the microcoded control:

$$C_{micro} = C_{ff}(\zeta) + C_{Inc}(\zeta) + C_{Select} + \zeta \cdot C_{and} + C_{inv}$$

$$+ \begin{cases} C_{ROM}(k, \alpha) + C_{ff}(\alpha) & \text{, unary} \\ C_{ROM}(k, a) + C_{ff}(\alpha) + C_{\mu Dec} & \text{, binary, in front} \\ C_{ROM}(k, a) + C_{ff}(a) + C_{\mu Dec} & \text{, binary, behind} \end{cases}$$

$$A_{CON}(Csig) = \begin{cases} D_{\mu Dec} & \text{, binary, behind } \mu\text{-IR} \\ 0 & \text{, otherwise} \end{cases}$$

$$T_{micro} = \max\{D_{Inc}(\zeta), A(In)\} + D_{Select} + D_{and} + \Delta$$

$$+ \begin{cases} D_{ROM}(k, \alpha) & \text{, unary} \\ D_{ROM}(k, a) + D_{\mu Dec} & \text{, binary, in front } \mu\text{-IR} \\ D_{ROM}(k, a) & \text{, binary, behind } \mu\text{-IR} \end{cases}$$

It is cheaper to place the decoders behind the microinstruction register, because this register then needs to be less wide, but the decoders might slow down the cycle time of the whole machine. However, *encoding the control signals* usually reduces the cost of the microcoded control without increasing the cycle time by much.

### Hardwired Control Assistance

Several control signals are necessary to specify the operation of units like the ALU, the shifter or the memory, or to specify the correct register. Devices for microcoded controls often allow this information to be taken directly from the macro-instruction word reducing the width of the $\mu$-ROM. For example, on a load, the microinstruction word provides the memory read signal, but the macro-instruction word specifies the width of the memory access.

The direct use of bits from the macro-instruction enables the microcoded control to execute several machine instructions using exactly the same microcode. This optimization consequently reduces the width and the length of the $\mu$-ROM and should be used whenever possible.

| $\mu type[1:0]$ | address calculation |
|:---:|:---:|
| 00 | branch |
| 01 | loop |
| 10 | table lookup |

Table 9.1: Coding of the microinstruction type

## 9.2 Microcoded DLX Architecture

The hardwired control automaton of the DLX architecture can certainly be replaced by a microcoded control, but the profitability is another matter. Since both controls implement the same FSD (figure 8.8), this modification has no direct impact on the rest of the design. The cost and cycle time of the microcoded DLX design will be different, but the CPI value remains the same.

According to figure 9.1, the microcoded control unit comprises the micro-PC, an incrementer, the micro-IR, the environment of the microprogram ROM and the address selection unit. In order to evaluate the DLX designs with a microcoded control unit, we specify the microinstructions of the DLX, derive an implementation of the address selection unit (section 9.2.2) and describe the modifications due to encoding of the microinstructions (section 9.2.3).

### 9.2.1 The Microinstruction Word

As in the general case, the microinstructions of the DLX machine comprise the three fields $\mu cond$, $\mu adr$ and $\mu sig$.

#### Format and Semantics of $\mu cond$

The microcondition $\mu cond$ codes the transition function of the DLX control automaton. The transitions fall under one of three types — branch, loop, table lookup — and are coded as indicated in table 9.1. We deal separately with microbranches, microloops, table lookups and finally interrupts.

In section 8.1.4, we specified the transition function of the DLX control automaton by the FSD of figure 9.4. Most states of this FSD only have one direct successor. Thus, the transition of these states can be handled by incrementing the micro-PC or by an unconditional branch.

The states AdrComp, SavePC and Branch have an outdegree of two. In all three cases, the microcode of the state and of one of its successors can be stored consecutively. Thus, the target address can be specified by a conditional microbranch. The branch condition of these branches can be derived from table 9.2 which is an excerpt of table 8.9. Thus, for the microbranches we need 5 microbranch conditions: the three conditions from table 9.2 and the two conditions 1 and 0 for increments and gotos.

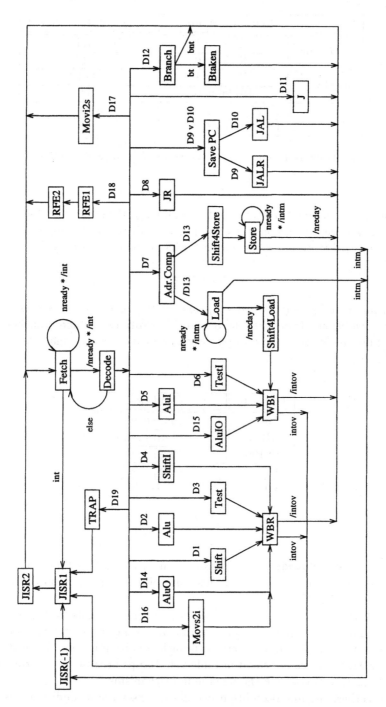

Figure 9.4: FSD of the DLX design supporting interrupts ($k = 32$ states)

| State $z$ | Relevant Inputs $I$ | Target State $next_z(I)$ |
|---|---|---|
| AdrComp | $IR[29] = 1$ | Shift4Store |
|  | $IR[29] = 0$ | Load |
| SavePC | $IR[28] = 1$ | JALR |
|  | $IR[28] = 0$ | JAL |
| Branch | $IR[26] \otimes AEQZ = 1$ | Btaken |
|  | $IR[26] \otimes AEQZ = 0$ | Fetch |

Table 9.2: Transition function for the states `AdrComp`, `SavePC` and `Branch`

If we ignore jumps to states `JISR1` or `JISR(-1)`, then the states `Fetch`, `Load` and `Store` have a common transition scheme, which can be implemented by a microloop. All these loops depend on the same loop condition

$$nready = 1 \ .$$

Thus, for microloops (specified by $\mu type = 01$), there is no need to specify a branch condition. Therefore, the 5 branch conditions of microbranches are all the conditions we need. For a first design, we code them in unary as indicated in table 9.3.

| $\mu bc[4:0]$ | Condition |
|---|---|
| 10000 | 0 (increment) |
| 01000 | 1 (goto $\mu adr$) |
| 00100 | $IR[29] = 1$ ? |
| 00010 | $IR[28] = 1$ ? |
| 00001 | $IR[26] \otimes AEQZ = 0$ ? |

Table 9.3: Coding of the microbranch condition

For the state `Decode`, we compute the next state by a table lookup. The next state function $next_{decode}$ for this state properly depends on the 12 inputs IR[31 : 26] and IR[5 : 0].

The transition function of the FSD in figure 9.4 depends on three interrupt signals. For each of these signals, we reserve an interrupt branch condition as indicated in table 9.4. If an interrupt branch condition is true, the next state is `JISR` or `JISR(-1)` as indicated in table 9.4. Jumps because of interrupt branch conditions simply override the remaining mechanisms.

Altogether, the microcondition $\mu cond$ comprises the 3 condition bits $ibc$ for interrupts, the 5 bits $\mu bc$ for microbranch conditions and the 2 bits $\mu type$ for the type of the microinstruction. Thus, the unary coded micro-condition of the DLX machine is $\beta = 10$ bits wide.

| $ibc[2:0]$ | Branch Condition | Destination |
|:---:|:---:|:---:|
| 100 | $int = 1$ ? | JISR1 |
| 010 | $intov = 1$ ? | JISR1 |
| 001 | $intm = 1$? | JISR(-1) |

Table 9.4: Coding of the interrupt branch condition

### Format of the Field $\mu sig$

In a first design, we use unary coding of the output signals $Csig[1:\gamma]$, i.e., for each $i \in \{1, \ldots, \gamma\}$, we reserve bit $\mu sig[i]$ for output signal $Csig[i]$. For each instruction, the values of these bits can immediately be read off from the tables 6.13 and 8.7.

### The Microprogram of the DLX Machine

Table 9.5 lists the whole microprogram of the DLX machine without the field $\mu sig$. The table also indicates how the microinstructions correspond to the states of the FSD.

Thus, the microprogram of the DLX machine requires $k = 32$ lines of code, and the microprogram ROM can be addressed by $\zeta = 5$ bits. Under unary coding of the control signals and of the microcondition, the microinstructions of the DLX machine are

$$\alpha = \zeta + \beta + \gamma = 5 + 10 + 39 = 54$$

bits wide.

## 9.2.2   Address Selection Unit

In this section, we adapt the address selection unit of figure 9.2 to the requirements of the DLX machine. The specification of the microinstructions (section 9.2.1) indicates that the address selection unit has to handle microbranches, one microloop, one lookup table and two addresses for interrupt microcode. We therefore implement the address selection unit of the microcoded DLX control as indicated in figure 9.5.

### Generating the Select Signals

The multiplexer with select signal $bc$ implements the conditional microbranch with the five branch conditions indicated in table 9.3. Signal $bc$ can therefore be generated as

$$bc = (\mu bc[3] \vee \mu bc[2] \wedge IR[29] \vee \mu bc[1] \wedge IR[28])$$
$$\vee \mu bc[0] \wedge (\overline{IR[26] \otimes AEQZ}))$$

| μ-ROM Address | Microcondition μcond | | | μadr | Corresponding State |
|---|---|---|---|---|---|
| | ibc | μtype | μbc | | |
| 0 | 100 | 01 | ***** | Decode | Fetch |
| 1 | 000 | 10 | ***** | ***** | Decode |
| 2-6 | 000 | 00 | 01000 | WBR | AluO, Alu, Shift, Test, ShiftI |
| 7 | 000 | 00 | 10000 | ***** | Moves2i |
| 8 | 001 | 00 | 01000 | Fetch | WBR |
| 9-11 | 000 | 00 | 01000 | WBI | AluIO, AluI, TestI |
| 12 | 000 | 00 | 10000 | ***** | Shift4Load |
| 13 | 001 | 00 | 01000 | Fetch | WBI |
| 14 | 000 | 00 | 00100 | Shift4Store | AdrComp |
| 15 | 010 | 01 | ***** | Shift4Load | Load |
| 16 | 000 | 00 | 10000 | ***** | Shift4Store |
| 17 | 010 | 01 | ***** | Fetch | Store |
| 18 | 000 | 00 | 00010 | JALR | SavePC |
| 19-20 | 000 | 00 | 01000 | Fetch | JAL, JALR |
| 21 | 000 | 00 | 00001 | Fetch | Branch |
| 22-25 | 000 | 00 | 01000 | Fetch | Btaken, JR, J, Movei2s |
| 26 | 000 | 00 | 01000 | JISR1 | Trap |
| 27 | 000 | 00 | 10000 | ***** | REF1 |
| 28 | 000 | 00 | 01000 | Fetch | REF2 |
| 29-30 | 000 | 00 | 10000 | ***** | JISR(-1), JISR1 |
| 31 | 000 | 00 | 01000 | Fetch | JISR2 |

Table 9.5: Microprogram of the DLX machine

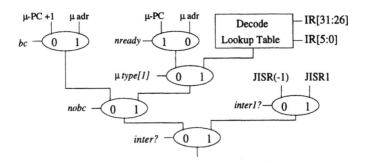

Figure 9.5: Address selection of the microcoded DLX control. JISR1 and JISR(-1) indicate the address of the corresponding microinstructions

The multiplexer with select signal *nready* implements the single cycle loop of a memory access ($\mu type = 01$). On an active signal *nready*, the control returns to the current microinstruction, otherwise it jumps to the address $\mu adr$.

According to the definition of the microinstruction type (table 9.1), the bit $\mu type[1]$ is only active on a table lookup. Thus, this flag selects between the table lookup and the loop mechanism. Since on a microbranch, neither of the $\mu type$ flags is active, the select signal *nobc* (no branch) can be generated as

$$nobc = \mu type[1] \lor \mu type[0].$$

As stated in table 9.4, the DLX microcontrol handles three interrupt branch signals. These conditions specify, whether the jump to the ISR starts in state JISR1 or in state JISR(-1). Signal *inter1?* selects between the two entree points, and signal *inter?* indicates that at least one of the interrupt branch conditions is true. These two select signals are generated as

$$inter1? = ibc[2] \land int \lor ibc[1] \land intov$$
$$inter? = ibc[2] \land int \lor ibc[1] \land intov \lor ibc[0] \land intm.$$

### The Decode Lookup Table

During decode, the microcoded control has to branch to one out of 17 states based on the two opcodes (12 bits). At the first view, these transitions seem to depend properly on both opcodes. Thus, the lookup mechanism introduced in section 9.1.2 requires a $2^{12} \times 5$ bit ROM for the lookup table.

However, the coding of the DLX instruction set (tables 6.1, 6.2 and 6.3) indicates, that only R-type instructions specify a secondary opcode IR[5:0] and that for these instructions the primary opcode IR[31:26] equals zero. Consequently, the 7 bits

$$tadr[6:0] = \begin{cases} 1\,IR[5:0] & \text{, for R-type, i.e., } IR[31:26] = 0^6 \\ 0\,IR[31:26] & \text{, otherwise} \end{cases}$$

are sufficient for the lookup in the decode table. The circuit of figure 9.6 implements this optimized version of the DLX lookup table. Since all the inputs of the decode table are provided directly by registers, the cost and the accumulated delay of the lookup table can be expressed as:

$$C_{DecTab} = \begin{cases} C_{ROM}(2^{12}, 5) & \text{, full table} \\ C_{ROM}(2^7, 5) + C_{zero}(6) + C_{mux}(6) & \text{, optim. table} \end{cases}$$

$$A_{DecTab} = \begin{cases} D_{ROM}(2^{12}, 5) & \text{, full table} \\ D_{ROM}(2^7, 5) + D_{zero}(6) + D_{mux}(6) & \text{, optimized table} \end{cases}$$

### Cost and Delay of the Address Selection Unit

The microprogram ROM with its $k = 32$ lines of code is addressed by $\zeta = 5$ bits wide addresses. Thus, the multiplexers of the address selection unit are also 5

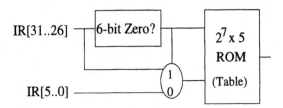

Figure 9.6: Improved implementation of the instruction decode

bits wide, and the address selection circuit of figure 9.5 has cost

$$C_{AdrSel} = C_{DecTab} + 6 \cdot C_{mux}(5) + 6 \cdot C_{and} + 6 \cdot C_{or} + C_{xnor}.$$

Let the microcondition and the control signals of the macro-machine be valid $A_{CON}(Csig)$ delays after the start of the cycle. As stated in section 8.2.2 (page 170), the inputs $AEQZ$, $nready$ and the interrupt signals $int$, $intm$, $intov$ have the accumulated delays $A_{GPR}(AEQZ)$, $A_{Menv}(nready)$ respectively $A_{CaPro}$. Thus, the output of the address selection unit is then valid $A_{AdrSel}$ delays after the start of the cycle, where

$$
\begin{aligned}
A_{AdrSel} = \quad & \max\{A_{DecTab} + 3 \cdot D_{mux}(5), & \text{[table]} \\
& A_{CON}(Csig) + 4 \cdot D_{mux}(5), & \text{[$\mu$adr]} \\
& D_{Inc}(5) + 3 \cdot D_{mux}(5), & \text{[$\mu PC + 1$]} \\
& \max\{A_{CON}(Csig), A_{CaPro}\} + D_{and} + D_{or} \\
& \quad + \max\{D_{mux}(5), D_{or}\} + D_{mux}(5), & \text{[JISR]} \\
& A_{Menv}(nready) + 4 \cdot D_{mux}(5), & \text{[$nready$]} \\
& A_{CON}(Csig) + D_{and} + 3 \cdot D_{or} + 3 \cdot D_{mux}(5), & \text{[$nobc, bc$]} \\
& A_{GPR}(AEQZ) + D_{xnor} + D_{and} + D_{or} \\
& +3 \cdot D_{mux}(5)\} & \text{[$AEQZ$]}
\end{aligned}
$$

### Illegal Instruction Detection

The address of the next microinstruction is also relevant for the interrupt unit to detect an illegal macro-instruction. For that purpose, the DLX with hardwired control tests for a transition from **Decode** to **Fetch** (section 8.1.4) and indicates such a transition by interrupt event signal env[1].

The microcoded control must generate this event signal too. For that purpose, it must detect whether the current microinstruction equals **Decode** and whether the target address of the address selection circuit equals zero (address of the fetch microcode). According to the DLX microprogram (table 9.5), flag $\mu type[1]$ is only active during the decode instruction. Furthermore, none of the interrupt branch condition flags $ibc$ are active. In the decode cycle, the result of the table lookup is therefore forwarded to the output of the address selection unit. Thus,

| Field | Meaning | unary | binary |
|-------|---------|-------|--------|
| $ibc[2:0]$ | interrupt flags | 3 | 2 |
| $\mu type[1:0]$ | microinstruction type | 7 | 3 |
| $\mu bc[4:0]$ | microbranch condition | | |
| Width of the microcondition $\mu cond$ | | $\beta = 10$ | $b = 5$ |

Table 9.6: Grouping and encoding of microcondition

the microcoded DLX control generates the interrupt event signal evn[1] as

$$evn[1] = \mu type[1] \wedge (output(\text{decode table}) = 0^5).$$

Let the microcondition be valid $A_{CON}(Csig)$ delays past the start of the cycle. An illegal instruction can then be detected at the following cost and delay:

$$
\begin{aligned}
C_{ILL} &= C_{zero}(5) + C_{and} \\
A_{ILL} &= \max\{A_{CON}(Csig), A_{DecTab} + D_{zero}(5)\} + D_{and}.
\end{aligned}
$$

### 9.2.3  Encoded Control Signals

The DLX microinstruction format of section 9.2.1 codes the microcondition and the control signals of the macro-machine in unary, but both groups of signals have a big potential for encoding.

#### Encoding the Microcondition

The microbranch condition $\mu bc$ is only used during microbranches ($\mu type = 00$). Thus, at most one of the 5 flags $\mu bc$ and of the 2 flags $\mu type$ for the type of the microinstrution can be active per microinstruction. These 7 signals can therefore be mapped into the same group. The 3 interrupt flags must be considered separately; but both groups of signals can be coded in binary. According to table 9.6, encoding reduces the width of the microcondition $\mu cond$ from $\beta = 10$ to $b = 5$ bits.

#### Encoding the Macro-control Signals

The field $\mu sig$ controlling the macro-machine comprises six groups of related signals (figure 9.7). They specify two sources, a destination and a device which generates the result. They also control the functionality of the ALU and the shifter and indicate special events (Misc). 

The enable signals of tristate drivers connected to the same bus can be binary coded without any problem. For the remaining fields, we have to take precautions because several of their signals might be active during the same cycle. To circumvent those conflicts, we either have to split those fields or the

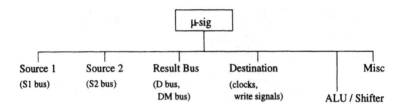

Figure 9.7: Partitioning of the microinstruction field $\mu$sig

| Field | Signals | unary | binary |
|-------|---------|-------|--------|
| Source 1 | AS1doe, BS1doe, PCS1doe, spr, MARS1doe, MDRS1doe | 6 | 3 |
| Source 2 | BS2doe, IRS2doe, 0S2doe, 4S2doe | 4 | 2 |
| Result | ALUzrDdoe, ALUzcDdoe, SHDdoe, MR, ESRDioe, EPCDioe, JISR2 | 7 | 3 |
| Destination | ABce, IRce, MARce, MDRce, RFw, SRice, EPCice | 7 | 3 |
|  | PCce, MW, spw, Cce | 4 | 2 |
| ALU/Shift | add, shift, ovf? | 3 | 2 |
| Misc | test, Jlink | 2 | 1 |
|  | fetch, ls, Itype, Jjump, trap, SRS12 | 6 | 3 |
| Width of the $\mu$-instruction field $\mu$sig | | $\gamma = 39$ | $g = 19$ |

Table 9.7: Grouping and encoding of the field $\mu$sig

critical microinstructions. Since the second solution increases the length of the microprograms; it requires additional states in the FSD and increases the CPI ratio of the design variant. We therefore prefer the first solution. Table 9.7 lists how the control signals are assigned to the fields and how they can be encoded. After encoding, the control signals only require $g = 19$ instead of $\gamma = 39$ bits in the microinstruction word.

### Impact on the Microcoded DLX Control

With unary coded microcondition and macro-control signals, the DLX microinstructions are $\alpha = \zeta + \beta + \gamma = 54$ bits wide. Encoding reduces its width to $a = \zeta + b + g = 29$, that is less than 54% of the original width. However, encoding also adds five 3-bit decoders, four 2-bit decoders and an 1-bit decoder with the following cost and delay:

$$C_{\mu Dec} = 5 \cdot C_{decf}(3) + 4 \cdot C_{decf}(2) + C_{decf}(1)$$
$$D_{\mu Dec} = D_{decf}(3).$$

As indicated in section 9.1.3, these decoders can be placed in front or behind the microinstruction register. Adapting the general cost and delay formulae of that section to the requirements of the DLX control automaton yields the following formulae. Note, that the outputs of the address selection unit are valid $A_{AdrSel}$ delays after the start of a new cycle. This includes the delay of the incrementer, the delay of the inputs of the control automaton and the delay $A_{CON}(Csig)$.

$$
\begin{aligned}
C_{auto} \;=\;& C_{ff}(\zeta) + C_{Inc}(\zeta) + \zeta \cdot C_{and} + C_{inv} + C_{AdrSel} \\
& + \begin{cases} C_{ROM}(k, \alpha) + C_{ff}(\alpha) & \text{, unary} \\ C_{ROM}(k, a) + C_{ff}(\alpha) + C_{\mu Dec} & \text{, binary, in front} \\ C_{ROM}(k, a) + C_{ff}(a) + C_{\mu Dec} & \text{, binary, behind} \end{cases}
\end{aligned}
$$

$$
A_{CON}(Csig) \;=\; \begin{cases} D_{\mu Dec} & \text{, binary, behind } \mu\text{-IR} \\ 0 & \text{, otherwise} \end{cases}
$$

$$
\begin{aligned}
T_{auto} \;=\;& A_{AdrSel} + D_{and} + \Delta \\
& + \begin{cases} D_{ROM}(k, \alpha) & \text{, unary} \\ D_{ROM}(k, a) + D_{\mu Dec} & \text{, binary, in front } \mu\text{-IR} \\ D_{ROM}(k, a) & \text{, binary, behind } \mu\text{-IR} \end{cases}
\end{aligned}
$$

### 9.2.4   Hardware Cost and Cycle Time

In the previous sections, we described some implementations of microcoded control and derived formulae for their cost and delay. The corresponding C routines are listed in appendix B.9. We provide two implementations of the decode lookup table and three versions for coding and decoding the microinstructions. That adds up to six DLX design variants with microcoded control. In this section, we will evaluate these six design variants and compare them with the DLX design with hardwired control.

### Cycle Time of the Microcoded DLX Design

The control automaton impacts the cycle time of the DLX in two ways. First, it generates signals $Csig$ to control the hardware. They are valid $A_{CON}(Csig)$ delays after a cycle started. Second, the control automaton also requires some time to update its register. This time is covered by $T_{auto}$. Table 9.8 lists these two delays and the cycle time of the cause environment and of the DLX architecture for all six control variants, assuming a memory status time of $d_{mstat} = 5$ gate delays.

The standard control signals $Csig$ are valid immediately at the beginning of a new cycle, for four of the six variants of the microcoded control. Only in the variants which store the encoded instruction word into register $\mu$-IR, the standard control signals first have to pass the decoder circuit. That causes a delay of 5 (3) gate delays. Since the longest paths through the data paths first

| | Implementation | $T_{auto}$ | | Delay | $T_{CAenv}$ | $T_{DLX}$ |
|---|---|---|---|---|---|---|
| | Decode Table | full | optim. | $Csig$ | | |
| | unary | 73 | 63 | 0 | 65 | 80 |
| Motorola | binary, pre | 77 | 67 | 0 | 65 | 80 |
| | binary, past | 72 | 62 | 5 | 70 | 85 |
| | unary | 38 | 35 | 0 | 40 | 49 |
| Venus | binary, pre | 41 | 38 | 0 | 40 | 49 |
| | binary, past | 38 | 35 | 3 | 43 | 52 |

Table 9.8: Delay of the microcoded control unit. Version 'binary, pre' decodes the encoded microinstruction before storing it in $\mu$-IR, and version 'binary, past' decodes it afterwards.

require some control signals, the accumulated delay of the control signals $Csig$ adds completely to the cycle time of the DLX data paths.

Optimizing the decode table improves the cycle time of the control automaton up to 14% (8% under Venus technology). Optimizing the microinstruction format (coding of the control signals and of the microcondition) has a slightly smaller impact (7%) on the cycle time of the control automaton. However, the table 9.8 indicates that even without these optimizations, the data paths requires more time than the automaton, at least for a memory status time of 5 gate delays.

The accumulated delay of input signal *nready* depends on the memory status time $d_{mstat}$. A parameter study indicates that the memory status time could even rise up to 45% (35%) of the cycle time of the data paths and the control automaton would still not slow down the DLX design. This is a similar result that rule 8.2 formulates for DLX with hardwired control. Thus

**Maxim 9.1 (Cycle Time of Microcoded Control)** *If the microcoded control of the non-pipelined DLX is realized as described above, and if the memory status time $d_{mstat}$ is small, the control does not lie on the time critical path of the design. However, decoding the encoded fields of the microinstruction word after storing it into register $\mu$-IR slows the cycle time down by about 6%.*

**Hardware Cost of the Microprogrammed DLX**

Table 9.9 lists the cost of the data paths, of the control unit and of the whole DLX design, assuming that the Motorola (Venus) technology provides ROMs at 1/8 (1/6) times the cost of equally sized RAMs (see section 2.4.1).

Optimizing the decode table shrinks the table from $2^{12}$ to $2^7$ lines of code. That reduces the cost of the microcoded control roughly by a factor of 5 (respectively 9) and reduces the cost of the whole DLX design by 32% (respectively 52%). Due to very expensive storage components, the Venus technology yields the bigger gains.

| Implementation: Decode Table | $C_{DP}$ any | $C_{CON}$ full | opt. | $C_{DLX}$ full | opt. |
|---|---|---|---|---|---|
| Motorola — unary | | 8856 | 1935 | 21953 | 15032 |
| Motorola — binary, pre | 13097 | 8814 | 1893 | 21911 | 14990 |
| Motorola — binary, past | | 8614 | 1693 | 21711 | 14790 |
| Venus — unary | | 45975 | 6318 | 77161 | 37504 |
| Venus — binary, pre | 31186 | 44551 | 4894 | 75737 | 36080 |
| Venus — binary, past | | 44251 | 4594 | 75437 | 35780 |

Table 9.9: Cost of the microcoded DLX. Version 'binary, pre', decodes the encoded microinstruction before storing it in $\mu$-IR, and version 'binary, past' decodes it afterwards.

Optimizing the coding of the microinstruction roughly halfs the width of the microprogram ROM. This also improves the cost of the control and the cost of the DLX machine, but at most by a factor of 28% respectively 5%. Thus:

**Maxim 9.2** *On a non-pipelined DLX design with microcoded control, a well adapted instruction format (respectively a well designed decode table) is essential for a microcoded control to be competitive. A naive approach may result in a huge cost increase.*

**Parameter Study**

The tables 9.8 and 9.9 also indicate, that only two out of the six design variants are reasonable, namely the two designs with optimized decode table and encoded microinstruction. Let design V1 decode the microinstruction before storing it in the register $\mu$-IR and let design V2 decode it afterwards.

Version V2 is about 6% slower and only slightly cheaper (2%) than version V1. It therefore depends on how much emphasis is put on the cost of the designs, whether version V1 or V2 is of higher quality. However, for a ROM scaling factor $ROMfac$ of 6 to 8, the quality gain of V1 over V2 stays under 6%.

Diagram 9.8 depicts the equal quality parameter $EQ(V1, V2)$ of both designs as a function of the factor $ROMfac$ by which a ROM is cheaper than a RAM. Since the EQ value is greater than 0.8, the cheaper of the two design variants (V2, decoder behind register $\mu$-IR) is only better if the cost of the design has a much stronger impact on the quality than the performance.

Thus, under a realistic quality metric $Q_q$ with $q \in [0.2, 0.5]$, only variant V1, which decodes the microinstruction before storing it in a register and which uses an optimized decode table, is a realistic design.

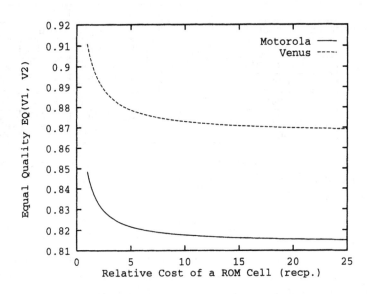

Figure 9.8: Equal quality value $EQ(V1, V2)$ of the microcoded DLX designs with optimized decode table and encoded microinstruction

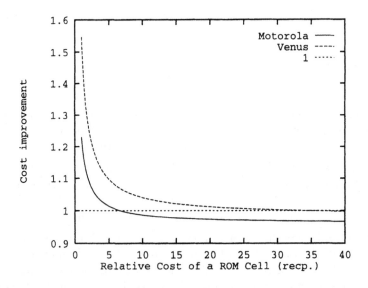

Figure 9.9: Cost improvement of a microcoded DLX over a DLX with hardwired control $(C_{DLX}(\mu\text{coded})/C_{DLX}(\text{hardwired}))$ as a function of the relative cost of ROM

**Comparison of Hardwired and Programmed Control**

The best DLX design variants with hardwired or microcoded control have the same cycle time. Differences in the quality are therefore exclusively caused by the cost of the control units. Diagram 9.9 depicts the cost ratio of the microprogrammed DLX and the hardwired DLX as a function of the factor $ROMfac$ by which a ROM is cheaper than an equally sized RAM.

Under Motorola technology, the break-even point lies around 7. A ROM cell then costs $C_{ROMcell} = C_{RAMcell}/7 = 0.29$ gate equivalents. Under Venus technology with the expensive storage components ($C_{RAMcell} = 12$), ROM must even be 37 times cheaper than RAMs before microcoded control becomes cheaper than hardwired control. A ROM cell then costs $12/37 = 0.32$ gate equivalents. Thus:

**Maxim 9.3** *For a non-pipelined DLX architecture, a design variant which uses a microcoded control can only compete with a design variant which uses a hardwired control, if the technology provides high density low cost ROMs, i.e., a ROM cell costs less than 0.35 gate equivalents.*

# Chapter 10

# Further Applications of the Architecture Model

Several Ph.D Theses used this method to analyze different topics of computer architecture.

For the first time, the model was applied in [Mül91] to attain quantitative analysis of several non-pipelined CPU designs, and to compare RISC and CISC approachs. Further analyses of CISC instruction set implementations were performed in [Sch92]. In his Thesis [Grü94], Grün compares I/O architectures and optimizes protocols.

The model was also used to evaluate supercomputer designs. In this field, Formella applied the model first [For92]. He used it to reverse engineer a CRAY-1 and to evaluate several vector computer designs. Based on these results, Massonne could show in [Mas94], that data flow machines can not compete with vector machines, at least under current technology constraints. The data flow principle itself turns out to be the real problem. In the PRAM project, Keller used the model to re-design the hardware simulation of a PRAM [Kel92]. This optimization made an efficient hardware simulation feasible.

In the most recent application of our model [Knu95], Robert Knuth studies the impact of pipelining on the design, cost and performance of the DLX architecture. That also includes the design of a precise interrupt mechanism. It turned out, that it is fairly straightforward, to adapt the proof of correctness from section 8.3 to the new interrupt mechanism.

# Appendix A

# Register Transfer Language

Usually, a register transfer language is used, to specify the meaning of machine instructions and to specify how they should be executed. In this book, we therefore use the following notations in order to describe how to manipulate the contents of a machine's registers and main memory.

| | |
|---|---|
| $=_n$ | indicates a data transfer or assignment of n-bit length. The subscript $n$ is omitted when the length is clear.<br>Rd = Rs    The contents of the source register Rs is transferred to the destination register Rd. |
| Rs$[m:n]$ | specifies a bit range of register Rs, to select parts of the register's contents. R1$[4:0]$ accesses the 5 least significant bits of register R1. |
| $X^n Y$ | concatenates the fields $X^n$ and $Y$. The superscript $n$ replicates the field $X$.<br>Rd = $0^{31}1$    The least significant bit of register Rd is set to 1 all other are cleared (0). |
| $\ll, \gg,$<br>$\ll_{arith},$<br>$\gg_{arith}$ | indicate logical left shift, logical right shift, arithmetic left shift and arithmetic right shift. The two left shifts are identical.<br>R1 $\ll 3$    shifts the contents of register R1 3 bits to the left. |
| $\wedge, \vee, \otimes$ | indicate the logical operations AND, OR and XOR. They are performed bitwise. |

M[ea]                    specifies a memory access. The effective address *ea*
                         selects a 1-byte data from memory.
                         $Rd = 0^{24}M[ea]$    reads a byte from memory, extends
                         it with zero and assigns it to register Rd.

( test ? arg1 : arg2 )
                         If the test is true, argument arg1 is taken otherwise
                         argument arg2 is taken.
                         $Rd = (Rs1 > Rs2 ? 1 : 0)$    If the contents of register
                         Rs1 is greater than the one of register Rs2 then Rd
                         is set to 1 and to 0 otherwise.

# Appendix B

# Cost and Run Time Programs

This chapter lists all the programs required to compute the cost and the cycle time of the designs described in this book. The programs are written in the programming language C and comprise several modules.

The module **evaluate** provides the quality ratio $Q_q(A, B,)$ and the equal quality parameter $EQ(A, B,)$ of section 2.3, and module **techno** handles the technology specific part of our architecture model, i.e., it defines the cost and the delay of the basic components of section 2.4.1. The modules **hardctr** and **microctr** provide the cost and the delay functions of the hardwired control respectively of the microcoded control. Module **macro** provides the combinatorial circuits of chapter 3 as a macro library of some sort. In order to refer to all these modules as a whole, we use the term *service modules*.

The architecture specific part is implemented in the main module which we call the *architectural module*. The architectural module comprises several parts. First, it specifies the cost and the delay of the different environments. It then provides functions for the cost of the data paths and of the whole machine, and it provides functions for the duration of all the cycles involved. Finally, the third part implements the evaluation of the design variants.

In order to enable a module $m$ to access the functions of a service module without any problems, each of the service modules consists of two files. The code file **file.c**, where file indicates the name of the service module, contains the actual C functions. The header file **file.h** specifies the types of the variables and functions and defines the data structures used.

All programs are available via FTP at **ftp-wjp.cs.uni-sb.de** in directory **/pub/architecture/{programs.tar.gz, programs.tar}** or via WWW at address **http://www-wjp.cs.uni-sb.de/teaching/books.html**.

# B.1    Module of the Quality Metric

In section 2.3, we introduced the quality ratio $Q_q(A, B, )$ of two designs and their
equal quality parameter $EQ(A, B, )$. These functions are required to evaluate
different variants of a design. We therefore added module **evaluate**.

## B.1.1    Header File of the Quality Functions

```
/*******************************************************************
**   evaluate.h       quality ratio Q(q,A,B) and equal quality  **
**                    parameter EQ(A,B) of two designs A,B       **
*******************************************************************/
/* q       : quality parameter q                                */
/* ca, ta : cost, delay of A       cb, tb : cost, delay of B    */
/* ------------------------------------------------------------- */
/* Let F be the faster and S be the slower of the designs.      */
/* EQ(F,S) = EQ(S,F) defined for                                */
/*   a) ts > tf  and  cs <= cf;   b) ts >= tf  and  cs < cf    */
/* Degenerated cases:                                           */
/*  ts = tf and cs = cf: always same quality:  EQ(S,F) := +2   */
/*  ts > tf and cs > cf: never same quality:   EQ(S,F) := -2   */
/* ------------------------------------------------------------- */

#ifndef EVALUATE_H
#define EVALUATE_H

extern double eval_Q();          /* q, ca, ta, cb, tb */
extern double eval_EQ();         /* ca, ta, cb, tb */

#endif
```

## B.1.2    Code File for the Quality Functions

```
/*******************************************************************
**   evaluate.c       quality ratio Q(q,A,B) and equal quality  **
**                    parameter EQ(A,B) of two designs A,B       **
*******************************************************************/

#include <math.h>

double eval_Q(q,ca,ta,cb,tb)
double q;
int ca,ta,cb,tb;
{ double c,t,r;

  r = (1.0*cb)/(1.0*ca);
  c = pow(r, q);
  r = (1.0*tb)/(1.0*ta);
  t = pow(r, 1.0-q);
  r = c*t;
```

```
  return r;
}

double eval_EQ(ca,ta,cb,tb)
int ca,ta,cb,tb;
{ double c,t,r;

  c = (1.0*ca) - (1.0*cb);
  t = (1.0*ta) - (1.0*tb);
  r = c * t;

  if ( r > 0)
  { /* faster design is cheaper */
    return -2;
  }
  else
  { if ((c == 0) && (t == 0))
    { /* both designs of equal quality */
      return +2;
    }
    else
    { /* standard case */
      c = (1.0*ca)/(1.0*cb);
      t = (1.0*ta)/(1.0*tb);
      r = log(t)/log(t/c);
      return r;
    }
  }
}
```

# B.2   Technology Specification

Cost and delay of the basic components (section 2.4.1) are the only technology
specific parameters. Here, we use two sets of technology parameters, described
by the following two files. One set indicated as *Motorola technology* is based on
Motorola's H4C CMOS sea-of-gate design series [NB93]. The other set is based
on the VENUS design system [HNS86, Sie88].

## B.2.1   Header File of the Technology Module

```
/*****************************************************************
**   techno.h              technology header module           **
*****************************************************************/

#ifndef TECHNO_H
#define TECHNO_H

#define MOTOROLA   0     /* Motorola H4C technology, 1993 */
#define VENUS      1     /* Venus Design System,
```

                        (Arno, SMM PhD Thesis) */

```
extern int cinv, cand, cor, cnand, cnor;
extern int cxor, cxnor, cmux, cdriv, cff;
extern int cramcell, delta;
extern double romfac;
extern int dinv, dand, dor, dnand, dnor;
extern int dxor, dxnor, dmux, ddriv, dff;

extern int cram(),  cram2(),  crom();
extern int dram(),  dram2(),  drom();

extern int   init_techno();      /* initialize technology
                                    parameters */
extern char* name_techno();      /* name: current technology */

#endif
```

## B.2.2   Code File of the Technology Module

```
/***************************************************************
** techno.c                   technology parameter module    **
***************************************************************/

#define IN_TECHNO
#include <stdio.h>
#include <math.h>
#include "techno.h"

extern int ld();
static int technology=0;

int cinv, cand, cor, cnand, cnor;
int cxor, cxnor, cmux, cdriv, cff;
int cramcell, delta;
double romfac;
int dinv, dand, dor, dnand, dnor;
int dxor, dxnor, dmux, ddriv, dff;

char* name_techno()                /* name of current technology */
{
  switch(technology)
  { case MOTOROLA:
      return (char*)"Motorola";
      break;
    case VENUS:
      return (char*)"Venus";
      break;
    default:
```

```
        return (char*)"unknown";
        break;
    }
}

int init_techno(par)        /* initialize cost and delay of basic
                               components */
int par;
{
  switch(par)
  { case MOTOROLA:
      cinv=1;  cand=2;  cor =2;   cnand=2;  cnor=2;
      cxor=4;  cxnor=4; cmux=3;   cdriv=5;  cff=8;
      cramcell=2;        romfac=8.0;          delta=1;
      dinv=1;  dand=2;  dor =2;   dnand=1;  dnor=1;
      dxor=2;  dxnor=2; dmux=2;   ddriv=2;  dff=4;
      break;
    case VENUS:
      cinv=1;  cand=2;  cor =2;   cnand=2;  cnor=2;
      cxor=6;  cxnor=6; cmux=3;   cdriv=6;  cff=12;
      cramcell=12;       romfac=6.0;          delta=1;
      dinv=1;  dand=1;  dor =1;   dnand=1;  dnor=1;
      dxor=2;  dxnor=2; dmux=2;   ddriv=1;  dff=4;
      break;
    default:
      fprintf(stderr,
              "techno.c: init_techno() unknown parameter\n");
      exit(1);
  }
  technology=par;
  return 0;
}

/* ------------------------------------------------------------*/
/*      ROM and RAM units (A x n)                              */
/* ------------------------------------------------------------*/

int cram(A,n)               /* on chip static RAM */
int A,n;
{ switch(technology)
  { case MOTOROLA:
      if (n>1)
        return (A+3)*(ld(ld(n))+n)*cramcell ;
      else
        return (A+3)*n*cramcell ;
      break;
    case VENUS:
      return A*n*cramcell ;
      break;
    default:
```

```
        fprintf(stderr,"techno.c: cram() unknown technology\n");
        exit(1);
  }
}

int dram(A,n) int A,n;
{ switch(technology)
  { case MOTOROLA:
        if (A >= 128)
          return 10 + 3*ld(A);
        else
          return ld(n) + A/4;
        break;
    case VENUS:
      return ceil(1.5*ld(A));
      break;
    default:
      fprintf(stderr,"techno.c: dram() unknown technology\n");
      exit(1);
  }
}


int cram2(A,n)                      /* Dualport RAM;  A < 128 */
int A,n;
{ switch(technology)
  { case MOTOROLA:
    case VENUS:
     return ceil(cram(A,n)*1.6);
     break;
    default:
     fprintf(stderr,"techno.c: cram2() unknown technology\n");
     exit(1);
  }
}

int dram2(A,n) int A,n;
{ switch(technology)
  { case MOTOROLA:
    case VENUS:
      return ceil(dram(A,n)*1.5);
      break;
    default:
      fprintf(stderr,"techno.c: dram2() unknown technology\n");
      exit(1);
  }
}

int crom(A,n)                       /* on chip ROM */
int A,n;
```

```
{return (int)(cram(A,n)/romfac);
}

int drom(A,n)
int A,n;
{return dram(A,n) ;
}
```

# B.3   Library of Functional Circuits

In chapter 3, we specified some combinatorial circuits, which are repeatedly used
in all our designs. These include decoders, an encoder, a shifter, different kinds
of adders and some other useful functional circuits. Module macro implements
their cost and delay formulae.

## B.3.1   Header File of the Macro Library

```
/*****************************************************************
** macro.h          commonly used functional modules         **
*****************************************************************/

#ifndef MACRO_H
#define MACRO_H

extern int max();       /* max of two integers */
extern int ld();        /* log base 2, rounded to next integer */
extern int power2();    /* 2^n */

extern int cdecs(), ddecs();    /* decoder, slow */
extern int cdecf(), ddecf();    /* decoder, fast */
extern int chdec(), dhdec();    /* half decoder */
extern int cenf(),  denf();     /* encoder, with flag */
extern int cenc(),  denc();     /* encoder, no flag */

extern int czero(), dzero();    /* zero tester */
extern int cinc(),  dinc();     /* incrementer, ripple carry */
extern int ccls(),  dcls();     /* cyclic left shifter */

extern int cFA(),   dFA();      /* full adder */
extern int cRCA(),  dRCA();     /* adder (carry ripple) */
extern int cPP(),   dPP();      /* parallel prefix */
extern int cCLA(),  dCLA();     /* adder (carry look ahead) */
extern int cCSA(),  dCSA();     /* adder (conditional sum) */

#endif
```

## B.3.2   Code File of the Macro Library

```
/*****************************************************************
** macro.c          commonly used functional modules         **
*****************************************************************/

#include "techno.h"     /* extern decl. of basic components */

/* === Some useful functions ================================ */

int max(a, b)           /* maximum of two integers */
```

```
int a, b;
{ return (a>b) ? a:b ;
}

int ld(n)              /* log base 2, rounded to the next integer */
int n;
{ return (n==1) ? 0:ld( (n+1)/2 ) + 1;
}

int power2(n)          /* 2^n */
int n;
{ return (n==0) ? 1:power2( n-1 )*2 ;
}

/* === Functional Modules (Macros) =========================== */

/* --- Cost and delay of n-bit decoders: n -> 2^n ----------- */

int cdecs(n)           /* simple decoder, but slow */
int n;
{ return (n==1) ? cinv : cdecs(n-1) + power2(n)*cand + cinv;
}

int ddecs(n)
int n;
{ return (n==1) ? dinv : max(ddecs(n-1),dinv) + dand;
}


int cdecf(n)           /* fast decoder */
int n;
{ return (n==1) ? cinv :
                cdecf(n/2) + cdecf((n+1)/2) + power2(n)*cand;
}

int ddecf(n)
int n;
{ return (n==1) ? dinv : ddecf((n+1)/2) + dand;
}

/* --- Cost and delay of n-bit half-decoder: n -> 2^n ------- */

int chdec(n)
int n;
{ return (n==1) ? 0 : chdec(n-1) + power2(n-1)*(cand + cor);
}

int dhdec(n)
int n;
{ return (n==1) ? 0 : dhdec(n-1) + max(dand, dor);
```

```
}

/* --- Cost, delay of n-bit encoder 2^n -> n   (with flag) --- */

int cenf(n)
int n;
{ return (n==1) ? cor : 2* cenf(n-1) + n*cor;
}

int denf(n)
int n;
{ return (n==1) ? dor : denf(n-1) + dor;
}

/* --- Cost, delay of n-bit encoder 2^n -> n   (no flag) ----- */

int cenc(n)
int n;
{ return (n==1) ? 0 : cenf(n-1) + cenc(n-1) + (n-1)*cor;
}

int denc(n)
int n;
{ return (n==1) ? 0 : max(denf(n-1), denc(n-1)) + dor;
}

/* --- Cost and delay n-bit Zero Test --------------------- */

int czero(n)
int n;
{ return (n-2)*cor + cnor;
}

int dzero(n)
int n;
{ return  ld(n/2)*dor + dnor;
}

/* --- Cost and delay n-bit Ripple Carry Incrementer -------- */

int cinc(n)
int n;
{ return (n-1)*(cand + cxor) + cinv;
}

int dinc(n)
int n;
{ return (n==1) ? dinv :
                max(dinv, max((n-1)*dand, (n-2)*dand +dxor));
}
```

```
/* --- Cyclic left shifter ------------------------------------ */
int ccls(n)
int n;
{ return ld(n)*n*cmux;
}

int dcls(n)
int n;
{ return ld(n)*dmux;
}


/* === different adder types (RCA, CLA, CSA) =============== */

/* --- Cost and delay of Parallel Prefix, n power of 2 ------ */

int cPP(n)
int n;
{ return (n==1) ? 0 : cPP(n/2) + n - 1;
}

int dPP(n)
int n;
{ return (n==1) ? 0 : dPP(n/2) + 2;
}

/* --- Carry look-ahead adder, n power of 2 ----------------- */
int cCLA(n)
int n;
{ return cPP(n)*(2*cand +cor) + (2*n+2)*cxor + (n+1)*cand + cor;
}

int dCLA(n)
int n;
{ return dPP(n)*(dand + dor) + 2*dxor + dand + dor;
}

/* --- Full Adder ------------------------------------------- */
int cFA()
{ return 2*cxor + 2*cand + cor;
}

int dFAab_s()
{ return 2*dxor;
}

int dFAabc_c()
{ return dxor + dand + dor;
}
```

```
int dFAc_c()
{ return dand + dor;
}

int dFAc_cs()
{ return max(dxor, dand + dor);
}

int dFA()
{ return max( max( dFAab_s(), dFAabc_c()),
              max( dFAc_c(), dFAc_cs()));
}

/* --- Ripple Carry Adder  ----------------------------------*/

int cRCA(n)
int n;
{ return n*cFA() + 2*cxor;
}

int dRCA(n)
int n;
{ if (n==1)
    return dxor + max( max(dFAabc_c(), dFAab_s()), dFAc_cs());
  else
    return dxor + max(dFAab_s(),
                      dFAabc_c() + (n-2)*dFAc_c() + dFAc_cs());
}

/* --- A2(n) adder ------------------------------------------ */

int cA2(n)
int n;
{ if (n==1)
    return cxor + cxnor + cand + cor ;
  else
    return cA2(n/2) + cA2((n+1)/2) + 2*(n/2+1)* cmux;
}

int dA2(n)
int n;
{ if (n==1)
    return max( max(dxor, dxnor), max(dand, dor)) ;
  else
    return dA2((n+1)/2) + dmux;
}

/* --- Conditions Sum Adder ------------------------------- */
```

```
int cCSA(n)
int n;
{ return  cA2(n) + (n+1)*cmux + 2*cxor;
}

int dCSA(n)
int n;
{ return dA2(n) + dmux + 2*dxor;
}
```

## B.4   Hardwired Control

In chapter 4, we derived cost and run time formulae for a Moore and a Mealy
implementation of a hardwired control automaton. The states can be coded in
binary or in unary. Module **hardctr** provides the corresponding C routines of
all four versions.

### B.4.1   Header File of the Hardwired Control

```
/**************************************************************
** hardctr.h       hardwired control logic module          **
**   Mealy and Moore implementations of the control automaton **
**************************************************************/

#ifndef CONTROL_H
#define CONTROL_H

extern int set_contype();       /* set ctrl automaton type */
                                /* par: mealy, binary */
extern int set_conpar();        /* set  automaton parameters */
                                /* states, inputs, outputs,
                                   #monomials, length max/sum,
                                   nu max/sum, fanin max/sum */
extern int aOut();              /* accum. delay of outputs */
/*      int aIN();              to be defined in machine spec */
extern int cauto();             /* cost of the automaton */
extern int tauto();             /* cycle time of the automaton */

                                /* circuit ILL detects illegal
                                   instructions */
extern int cILL();              /* cost of ILL */
extern int aILL();              /* accumulated delay of ILL */

extern void forall_line();      /* print horizontal line */
extern void forall_header();    /* print two line header */
extern void forall_print();     /* evaluate function and print */

struct control_parameters {
  int states, zeta;             /* #States, ld(states) */
  int sigma, gamma;             /* #input/output signals */
  int monom;                    /* #monomials m in state
                                   transitions */

  int lmax, lsum;               /* max, sum of literals in m */
  int numax, nusum;             /* max, sum of signal
                                   activations */

  int faninmax, faninsum;       /* max, sum fanin of states */
  int mealy, binary;            /* moore/mealy, unary/binary */
};

#endif CONTROL_H
```

## B.4.2  Code File of the Hardwired Control

```
/******************************************************************
** hardctr.c       hardwired control logic module            **
******************************************************************/

#include <math.h>
#include "techno.h"
#include "macro.h"
#include "hardctr.h"

static struct                        /* private control parameters */
        control_parameters con =
        {0,0, 0,0, 0,0,0,            /* must be set by set_con... */
         0,0, 0,0, 0,0};

int set_contype(m,b)                 /* set ctrl automaton type */
int m,b;                             /* mealy, binary */
{
  con.mealy = m;
  con.binary = b;
  return 0;
}

int set_conpar(s,i,o,m,mlm,mls,nm,ns,fm,fs)
int s,i,o,m,mlm,mls,nm,ns,fm,fs;
                                     /* set automaton parameters */
{
  con.states   = s;                 /* #states */
  con.zeta     = ld(s);
  con.sigma    = i;                 /* input/output signals */
  con.gamma    = o;
  con.monom    = m;                 /* monomials */
  con.lmax     = mlm;
  con.lsum     = mls;
  con.numax    = nm;                /* frequency of activity */
  con.nusum    = ns;
  con.faninmax = fm;                /* fanin */
  con.faninsum = fs;
  return 0;
}

/* === Cost and delay: Output circuit of the automaton ====== */

int c0()                    /* output circuit 0: Moore automaton */
{ return (con.nusum - con.gamma)*cor;
}
```

```
int d0()
{ return ld(con.numax) *dor;
}

int cOD()                   /* circuit OD (unary/binary) */
{ if (con.binary==0)
    return c0();
  else
    return cdecf(con.zeta) + c0();
}

int aOD()                   /* accumulated delay of circuit OD */
{ if (con.binary==0)
    return d0();
  else
    return ddecf(con.zeta) + d0();
}

int aOut()          /* accum. delay: outputs of the automaton */
{ if (con.mealy)
    return 0;
  else
    return aOD();
}

/* === Cost and delay: Next state circuit of the automaton == */

extern int aIn();       /* accumulated delay: inputs In */

/* --- Control Automaton ----------------------------------- */

int cCM()               /* Circuit CM (monomials) */
{ return con.sigma * cinv + (con.lsum - con.monom)* cand;}

int dCM()
{ return dinv + ld(con.lmax) *dand;}

int aCM()
{ return aIn() + dCM();}

int cCN()               /* Circuit CN (next state) */
{ return con.faninsum * (cand + cor) + (con.states - 1)* cor;}

int dCN()
{ return ld(con.faninmax) * dor + dand;}

int cNSE()              /* Circuit NSE (unary/binary) */
{ if (con.binary==0)
    return cCM() + cCN();
  else
```

```
     return cCM() + cCN() + cenc(con.zeta);
}

int aNSE()
{ int delay;

 if (con.binary==0)
   delay = aCM() + dCN() + dzero(con.states-1);
   else
   delay = max(aCM(), ddecf(con.zeta)) + dCN() + denc(con.zeta);

 if (con.mealy)  delay += dand;

 return delay;
}

int a0()                    /* for Mealy automata only */
{ int delay;

   delay = dand + max(dzero(con.states -1), d0()) + dor;

   if (con.binary==0)
    return aCM() + dCN() + delay;
   else
    return max(aCM(), ddecf(con.zeta)) + dCN() + delay;
}

/* === Cost of control automaton ============================ */

int cauto()
{ int cost;

   if (con.binary==0)
   { cost = cOD() +cNSE() +con.states *cff +czero(con.states-1)
          + cor + cinv + (con.states -1) * cand;

     if (con.mealy)
       cost += con.gamma * cff - cor;
   }
   else
   { cost = cOD() + cNSE() + con.zeta * (cff + cand) + cinv;

     if (con.mealy)
       cost += con.gamma * cff + czero(con.states -1)
              + (con.states - 1 - con.zeta) * cand;
   }

   return cost;
}
```

```
/* === cycle time of the control automaton =================== */

int tauto()
{ if (con.mealy)
    return max(aNSE(),  a0()) + dff + delta;
  else
  { if (con.binary==0)
      return aNSE() + max(dand, dor) + dff + delta;
    else
      return aNSE() + dand + dff + delta;
  }
}


/* === Testing for illegal instruction ==================== */
int cILL()
{ if ((con.binary==1)&&(con.mealy == 0))
    return czero(con.states -1) + cand;
  else
    return cand;
}

int aILL()
{ if (con.binary == 0)
    return aNSE() + dand;
  else
    return aNSE() + dand - denc(con.zeta) + dzero(con.states -1);
}


/* === Print functions ================================= */
void forall_line()
{
  printf(
  "-------------------------------------------------------\n");
  return;
}

void forall_header(str1,str2)
char *str1, *str2;
{
  printf("%-21s \t        Moore                Mealy\n",str1);
  printf("%-21s \t  unary  binary    unary  binary\n",str2);
  forall_line();
  return;
}

void forall_print(str,func)
char *str;
int (*func)();
```

```
{ printf("%20s:",str);
  set_contype(0, 0); printf("\t%7d",(*func)());
  set_contype(0, 1); printf("\t%7d",(*func)());
  set_contype(1, 0); printf("\t%7d",(*func)());
  set_contype(1, 1); printf("\t%7d\n",(*func)());
  return;
}
```

# B.5    MARK 1 like Designs

## B.5.1    Original MARK 1 Design

The following C-module is the architectural module of the original MARK 1
design introduced in chapter 5. The module also contains routines for the comparison of the different design variants.

```
/* ============================================================= */
/*  redesigned Mark 1 (hardwired control)                        */
/*  950817 TGR    1.0 - final version                            */
/* ============================================================= */

#include <stdio.h>
#include <math.h>
#include "techno.h"
#include "macro.h"
#include "hardctr.h"
#include "evaluate.h"

struct variable_parameters {
   int nd;                       /* data bus width */
   int na;                       /* address bus width */
   int dmem;                     /* memory access time */
   int adder_type;               /* choosen adder type */
} var;

/* ------------------------------------------------------------- */
/*      Environments (user defined macros)                       */
/* ------------------------------------------------------------- */

/* -- Composing the n-bit arithmetic unit AU (= subtractor) - */

int cau()
{ switch (var.adder_type) {
   case 0: return var.nd*cxor + cRCA(var.nd);
           break;
   case 1: return var.nd*cxor + cCLA(var.nd);
           break;
   case 2: return var.nd*cxor + cCSA(var.nd);
           break;
   default: fprintf(stderr,"cau(): undefined adder type\n");
           exit(1);
   }
}

int dau()
{ switch (var.adder_type) {
   case 0: return dxor + dRCA(var.nd);
           break;
   case 1: return dxor + dCLA(var.nd);
```

```
              break;
    case 2: return dxor + dCSA(var.nd);
              break;
    default: fprintf(stderr,"dau(): undefined adder type\n");
              exit(1);
    }
}

/* ------------------------------------------------------------ */
/*      calculation of the data path cost                       */
/* ------------------------------------------------------------ */

int caccenv()            /* register and check ACC<0 (no cost) */
{return var.nd*cff;}

int cirenv()             /* IR only */
{return (var.na+3)*cff;}

int cpcenv()             /* PC, adder, AMUX, PMUX and powerup */
{ int cadder;
  switch(var.adder_type) {
  case 0: cadder=cRCA(var.na); break;
  case 1: cadder=cCLA(power2(ld(var.na))); break;
  case 2: cadder=cCSA(power2(ld(var.na))); break;
  default: fprintf(stderr,"cpenv: undefined adder type\n");
          exit(1);
  }
  return var.na*cff+ cadder + 2*var.na*cmux
                  + var.na*cand + cinv + 2*cor;
}

int cmenv()              /* Memory and preceding MMUX */
{return var.na*cmux;}

/* --- Cost of the data path --------------------------------- */
int cDP()                /* all above parts + LMUX */
{return cau() + cirenv() + caccenv()
        + cpcenv() + cmenv() + var.nd*cmux;
}

/* ------------------------------------------------------------ */
/*      calculation of the cycle time                           */
/* ------------------------------------------------------------ */


int amadr()              /* memory addressing (PC, IR, MMUX) */
{ return dmux+max(aOut(),0); }


int amrd()               /* memory read (incl. MMux) */
```

```
{return var.dmem+max(amadr(),aOut());}


int tmemw()              /* memory write */
{return delta+var.dmem+max(amadr(),aOut());}


int tacc()               /* ACC */
{return (dff+delta)+dau()+max(amrd(),dmux + max(aOut(),0));}


int taddpc()             /* PC adder and AMUX */
{ int dadder;
  switch(var.adder_type) {
  case 0: dadder=dRCA(var.na); break;
  case 1: dadder=dCLA(power2(ld(var.na))); break;
  case 2: dadder=dCSA(power2(ld(var.na))); break;
  default: fprintf(stderr,"cpenv: undefined adder type\n");
          exit(1);
  }
  return dadder+dmux+max(aOut()+dor,amrd());}

int tpc()                /* PC */
{return (dff+delta)+dmux+max(aOut(),max(taddpc(),dand+amrd()));}

int tir()                /* IR */
{return (dff+delta)+max(aOut(),amrd());}

/* --- Cycle time of the data paths ------------------------- */
int tDP()
{return max(tmemw(),
       max(tacc(),
       max(tpc(), tir()
       )));
}

/* ----------------------------------------------------------- */
/*     Cost and cycle time of the Mark-1 Design              */
/* ----------------------------------------------------------- */

/* ----- Control Unit --------------------------------------- */

int aIn()                /* delay: inputs In of CON */
{return 0;
}

int cCONenv()
{ return cauto();
}
```

```
int tCON()
{ return tauto();
}

/* ----- whole machine ------------------------------------------ */

int cMark1()
{return cDP() + cCONenv();}

int tMark1()
{return max(tDP(), tCON());}

int cCONrel()
{ return (int)(100.*cCONenv()/cMark1());
}

/* ------------------------------------------------------------ */
/*      Evaluation of the design                                */
/* ------------------------------------------------------------ */

void main()
{
  char my_str[200];              /* temporary string */
  int i;                         /* loop variable */

  init_techno(MOTOROLA);         /* select technology */

                                 /* Control logic parameters */
  set_conpar(   9,               /* states of ctrl automaton */
                4, 8,            /* input/output of ctrl logic */
                7, 3, 21,        /* monomials (#, max, sum) */
                4, 12,           /* signal activity (max, sum) */
                2, 9);           /* fanin (max, sum) */

                                 /* design parameters */
  var.dmem = 20;                 /* main memory access time */
  var.nd   = 32;                 /* width of data bus */
  var.na   = 13;                 /* width of address bus */

  for(i=0;i<3;i++)
  { var.adder_type=i;    /* Select RCA, CLA, CSA */

    printf("\n\n\n Cost Delay Analysis of the Mark1 Design \n");
    printf(" Technology: %s, ", name_techno());
    switch(var.adder_type) {
      case 0: strcpy(my_str,"carry ripple"); break;
      case 1: strcpy(my_str,"carry look ahead"); break;
      case 2: strcpy(my_str,"conditional sum"); break;
      default: strcpy(my_str,"unknown");
    }
```

```
   printf(" Adder type:  %s\n", my_str);
   printf(" Dwidth = %d, Awidth= %d, Mem access time = %d\n",
          var.nd,var.na,var.dmem);

   forall_line();                     /* cost analysis */
   forall_header("cost","");
   forall_print("arithmetic unit",cau);
   forall_print("PC environment",cpcenv);
   forall_line();
   forall_print("data path",cDP);
   forall_print("control",cCONenv);
   forall_line();
   forall_print("Mark1",cMark1);
   forall_print("control/Mark1 [%]",cCONrel);

   forall_line();                     /* cycle time analysis */
   sprintf(my_str,"(dmem = %d)",var.dmem);
   forall_header("delay",my_str);
   forall_print("control signals",aOut);
   forall_print("memory read",amrd);
   forall_print("arithmetic unit",dau);
   forall_line();
   forall_print("memory write",tmemw);
   forall_print("accumulator",tacc);
   forall_print("instr. register",tir);
   forall_print("program counter",tpc);
   forall_line();
   forall_print("data path",tDP);
   forall_print("control",tCON);
   forall_line();
   forall_print("Mark1",tMark1);
 }

/* Quality ratios: (moore/bin), (moore/unary), (mealy/bin) */
{ double q;                           /* quality parameter */
  int dmem;                           /* memory access time */
  int cost_0, cost_1, cost_2, cost_n; /* cost */
  int time_0, time_1, time_2, time_n; /* time */
  double v1, v2, v3;                  /* temporary variables */
  FILE *fil;                          /* file struct */

  var.adder_type = 2;                 /* Choose CSA */

  fil=fopen("m1_q.dat","w");
  if(fil!=NULL)
    for(q=0.0; q<=1.0; q+=0.01)
    {
      set_contype(0,1); cost_n = cMark1(); time_n = tMark1();
      set_contype(0,0); cost_0 = cMark1(); time_0 = tMark1();
      set_contype(1,1); cost_1 = cMark1(); time_1 = tMark1();
```

```
          set_contype(1,0); cost_2 = cMark1(); time_2 = tMark1();

          v1=eval_Q(q, cost_0, time_0,
                       cost_n, time_n);
          v2=eval_Q(q, cost_1, time_1,
                       cost_n, time_n);
          v3=eval_Q(q, cost_2, time_2,
                       cost_n, time_n);

          fprintf(fil,"%6.2lf %8.5lf %8.5lf %8.5lf\n",
                     q, v1, v2, v3);
      }
    else
    { fprintf(stderr,"Could not open file 'm1_q.dat'\n");
      exit(1);
    }
    fclose(fil);
  }
}
```

## B.5.2   Modified Design M1_1

The following C-fragment lists the changes in the architectural description from the original MARK 1 design to the modified design M1_1.

```
/* ------------------------------------------------------------- */
/*      calculation of the data path cost                        */
/* ------------------------------------------------------------- */

int cacc()  /* register, test ACC<0 (no cost) */
{return var.nd*cff;}

int cir()  /* IR only */
{return (var.na+3)*cff;}

int cpcenv()  /* PC, incrementer, powerup and PMUX */
{
  return var.na*cff+ cinc(var.na)+
 var.na*cand+ cinv+ 2*cor+ var.na*cmux;
}

int cmenv() /* Memory and preceding MMUX */
{return var.na*cmux;}

/* --- Cost of the data path ------------------------------- */
int cDP() /* all above parts + LMUX */
{return cau() + cir() + cacc() + var.nd * cff
+ cpcenv() + cmenv() + var.nd*cmux;
}
```

```
/* ---------------------------------------------------------------- */
/*       calculation of the cycle time                              */
/* ---------------------------------------------------------------- */


int amadr() /* memory addr (PC, IR and MMUX) */
{ return dmux+max(aOut(),0); }


int amrd()  /* memory read (incl. MMux) */
{return var.dmem+max(amadr(),aOut());}

int tmdreg()
{return (delta+dff)+ max(amrd(), aOut());}


int tmemw() /* memory write */
{return delta+ var.dmem+max(amadr(),aOut());}


int tacc() /* ACC */
{return (delta+dff)+ dau()+max(0,dmux + max(aOut(),0));}


int aincpc() /* PC, incrementer and powerup logic */
{
   return dand+dinc(var.na);
}

int tpc() /* PC */
{return (delta+dff)+ dmux+max(aOut()+dor,max(aincpc(),0 ));}

int tir()   /* IR */
{return (delta+dff)+ max(aOut(),amrd());}

/* --- Cycle time of the datapath ---------------------------- */
int tDP()
{return max(tmemw(),
max(tmdreg(),
max(tacc(),
max(tpc(), tir()
))));
}
```

# B.6  DLX without Interrupt Handling

Module dlx1 is the architectural module of the DLX design of chapter 6, i.e., it specifies the cost and the delay of the DLX specific part of the design. The module also compares this design for the four implementations of the hardwired control.

The parameters of the struct var are implementation dependent. They include the width of the data paths, the number of general purpose registers and the access and status time of the main memory.

```
/* =============================================================== */
/* dlx1.c        architectural module    non-pipelined DLX       */
/*         no interrupt handling, hardwired control, CLA adder */
/* =============================================================== */

#include <stdio.h>
#include <math.h>
#include "techno.h"
#include "macro.h"
#include "hardctr.h"
#include "evaluate.h"

struct variable_parameters {
  int n;                         /* width of the data paths */
  int A;                         /* # GP-Registers */
  int dmem;                      /* memory access time */
  int dmstat;                    /* memory status time */
} var;

/* --------------------------------------------------------------- */
/*    Control Environment (1): Control Signals                     */
/* --------------------------------------------------------------- */
/* --- Delay: outputs Csig of the control automaton ---------- */
int aCON_Csig()
{ return aOut();
}

/* --- Memory Control circuit MC ---------------------------- */
int cMC()
{ return cdecf(2) + 16*cand + 10*cor + 3*cinv;
}

int aMC_R_Mw()
{ return max(ddecf(2), dinv + dand) + 2*dand + 2*dor;
}

int aMC_R_mis()
{ return max(dinv + 3*dand + 2*dor,
          ddecf(2) + 2*dor + 2*dand + max(dinv, dor));
}
```

```
int dMC_C_Mw()
{ return  dand + 2*dor;
}

int dMC_C_mis()
{ return dand;
}

int aCON_Mw()
{ return max(aCON_Csig() + dMC_C_Mw(), aMC_R_Mw());
}

int aCON_mis()
{ return max(aCON_Csig() + dMC_C_mis(), aMC_R_mis());
}

/* --- Control circuit PCC of the PC environment ------------- */
int cPCC()
{ return cor;
}

int dPCC()
{ return dor;
}

int aCON_PCce()
{ return aCON_Csig() + dPCC();
}

/* ------------------------------------------------------------ */
/*      ALU Environment (width: var.n)                          */
/* ------------------------------------------------------------ */
/* --- Arithmetic unit AU  (with carry look-ahead adder) ----- */
int cau()
{ return var.n * cxor + cCLA(var.n);
}

int dau()
{ return dxor + dCLA(var.n);
}

/* --- Comparator -------------------------------------------- */
int ccomp()
{ return czero(var.n) + 2*cinv + 4*cand + 2*cor;
}

int dcomp()
{ return max(dinv+dand, max(dzero(var.n)+dor, dzero(var.n)+dinv))
         + dand + dor;
```

```
}

/* --- ALU ----------------------------------------------------- */
int calu()
{ return  cau() +ccomp() +var.n *(cand +cor +cxor) +4*var.n*cmux;
}

int dALU_zc()                     /* delay of zc */
{ return dau() + dcomp();
}

int dALU_zr()                     /* delay of zr */
{ return max(dau() +dmux, max(dand, max(dor, dxor)) +3*dmux);
}

int dALU_ovf()                    /* delay of ovf */
{ return dau();
}

/* --- ALU Environment --------------------------------------- */
int caluenv()
{ return calu() + 2 *var.n *cdriv +2*3*cmux + cor;
}

int daluglue()                    /* delay of alu control */
{ return 2*dmux + dor;
}

int dalue_s_ZC()                            /* delay of ZC */
{ return dALU_zc() + ddriv;
}

int dalue_s_ZR()                            /* delay of ZR */
{ return dALU_zr() + ddriv;
}

int dalue_s_ovf()                           /* delay of ovf */
{ return dALU_ovf();
}

int dalue_c_ZC()                            /* delay of ZC */
{ return daluglue() + dALU_zc() + ddriv;
}

int dalue_c_ZR()                            /* delay of ZR */
{ return daluglue() + dALU_zr() + ddriv;
}

int dalue_c_ovf()                           /* delay of ovf */
{ return daluglue() + dALU_ovf();
```

```
}
/* ------------------------------------------------------------ */
/*     Shifter Environment                                      */
/* ------------------------------------------------------------ */
/* --- Distance computation (Dist) -------------------------- */
int cdist()
{ return cinc(5) + 6*cinv + 13*cmux;
}

int ddist()
{ return dinv + max(dinc(5), dmux) + 2*dmux;
}

/* --- Fill Bit --------------------------------------------- */
int cfill()
{ return 7*cmux + 2*cand + cinv;
}

int dfill()
{ return max(3*dmux, dinv) + dand + dmux;
}

/* --- Mask Generation (Mask) ------------------------------- */
int cmask()
{ return  chdec(5) + 4 * var.n * cmux + cor;
}

int dmask()
{ return max(dhdec(5), max(dmux, dor)) + 2*dmux;
}

/* --- Shift Unit ------------------------------------------- */
int csh()
{ return  ccls(var.n) +cdist() +cfill() +cmask() +var.n * cmux;
}

int dsh()
{ return max(ddist()+dcls(var.n), max(dfill(), dmask())) + dmux;
}

/* --- Shifter Environment ---------------------------------- */
int cshenv()
{ return csh() + var.n * cdriv;
}

int dshenv()
{ return dsh() + ddriv;
}
```

```
/* ----------------------------------------------------------- */
/*      Further DP Environments                                 */
/* ----------------------------------------------------------- */
/* --- IR Environment ---------------------------------------- */
int cirenv()
{ return 15*cmux +2*var.n*cdriv +var.n*cff;
}

int direnv()
{ return dmux + ddriv;
}

/* --- GPR Environment (# Registers: var.A, width: var.n) --- */
int cgpr()
{ return  cram2(var.A, var.n) + 3* var.n* cff + 3* var.n* cdriv
          + czero(var.n) + 2* czero(ld(var.A)) + 2* cinv
          + 2* ld(var.A)* cmux + 2* var.n* cand;
}

int dgpr_RF()
{ return 2*dmux + dram2(var.A, var.n);
}

int dgpr_dout()
{ return  max(dram2(var.A, var.n), dzero(ld(var.A)) +dinv)
          + dand;
}

int dgpr_Sb()
{ return ddriv;
}

int agpr_aeqz()
{ return dzero(var.n);
}

/* --- PC Environment ---------------------------------------- */
int cpcenv()
{ return 2 * var.n * cdriv + var.n * cff + var.n * cmux;
}

int dpcenv()
{ return ddriv;
}

/* --- Memory Environment ------------------------------------ */
int cmenv()
{ return 2 * var.n * cff + var.n * cmux + 2 * var.n * cdriv;
}
```

```
int dmenv_S1()
{ return ddriv;
}


/* ------------------------------------------------------------ */
/*      Cost and delay of the DLX Data Paths                    */
/* ------------------------------------------------------------ */
int cDP()
{ return caluenv() + cshenv() + cirenv() + cgpr()
        + cpcenv() + cmenv();
}

/* --- clocking of registers in the data paths --------------- */
int tRegce()
{ return max(aCON_Csig(), aCON_PCce()) + dff + delta;
}

/* --- register file write back ------------------------------ */
int tgpr__RF()
{ return aCON_Csig() + dgpr_RF() + delta;
}

/* --- register file read cycle ------------------------------ */
int tgpr_RF()
{ return aCON_Csig() + dgpr_dout() + dff + delta;
}

/* --- data on S1 and S2 bus --------------------------------- */
int aRegS()
{ return aCON_Csig() + max( max(direnv(), dpcenv()),
                       max(dgpr_Sb(), dmenv_S1()));
}

/* --- data from D bus into registers ------------------------ */
int dRegw()
{ return dmux;
}

/* --- shift cycles ------------------------------------------ */
int tSH()
{ return aRegS() + dshenv() + dRegw() + dff + delta;
}

/* --- ALU cycles -------------------------------------------- */
int talu_ZR()
{ return max(aCON_Csig() + dalue_c_ZR(), aRegS() + dalue_s_ZR())
        + dRegw() + dff + delta;
}
```

```
int talu_ZC()
{ return max(aCON_Csig() + dalue_c_ZC(), aRegS() + dalue_s_ZC())
        + dRegw() + dff + delta;
}

int aalu_ovf()
{ return max(aCON_Csig() + dalue_c_ovf(),
             aRegS() + dalue_s_ovf());
}

/* --- Cycle time of the data paths ------------------------- */
int tDP()
{ return max( max(tgpr__RF(), tgpr_RF()), max(tSH(),
             max( max(talu_ZR(), talu_ZC()), tRegce()))));
}

/* --- Memory cycle time tM --------------------------------- */
int tM()
{ return max(aRegS(), aCON_Mw()) +var.dmem +dRegw() +dff +delta;
}

int aMenv_mstat()
{ return max(aRegS(), aCON_Mw()) + var.dmstat;
}

/* ---------------------------------------------------------- */
/*   Control Environment (2): Next State Computation          */
/* ---------------------------------------------------------- */
/* --- accumulated delay: inputs In of control automaton ---- */
int aIn()
{ return max(agpr_aeqz(), aMenv_mstat());
}

/* ----- Control Unit --------------------------------------- */
int cCONenv()
{ return cauto() + cMC() + cPCC();
}

int tCON()
{ return tauto();
}

/* ---------------------------------------------------------- */
/*     Cost and cycle time of the DLX Design                 */
/* ---------------------------------------------------------- */
int cDLX()
{ return cDP() + cCONenv();
}

int tDLX()
```

```
{ return max(tDP(), tCON());
}

int cCONrel()
{ return (int)((100.0* cCONenv())/ cDLX());
}

int cGPRrel1()
{ return (int)((100.0* cgpr())/ cDP());
}

int cGPRrel2()
{ return (int)((100.0* cgpr())/ cDLX());
}

void set(ctr, techno)
int ctr, techno;
{ if (techno == 0)
    init_techno(MOTOROLA);
  else
    init_techno(VENUS);

  switch (ctr)
  { case 0:   set_contype(0, 0); return;
    case 1:   set_contype(0, 1); return;
    case 2:   set_contype(1, 0); return;
    case 3:   set_contype(1, 1); return;
    default: fprintf(stderr,
                "dlx1.c: set() unknown control type\n");
             exit(1);
  }
}

/* ------------------------------------------------------------ */
/*      Evaluation of the design                                */
/* ------------------------------------------------------------ */

void main()
{
  char my_str[200];              /* temporary strings */
  int i, x, y;                   /* loop variables */
  int c1, c2, t1, t2;            /* cost, time values */
  double res, q;                 /* result, quality parameter */

  init_techno(MOTOROLA);         /* select technology */

                                 /* Control logic parameters */
  set_conpar(  22,               /* states of state machine */
               14, 28,           /* input/output of control */
               22, 10, 97,       /* monomials (#, max, sum) */
```

```
                  10, 93,              /* signal activity (max, sum) */
                   4, 32);             /* fanin (max, sum) */

                                       /* design parameters */
  var.n        = 32;                   /* width of the data paths */
  var.A        = 32;                   /* number of GP-Registers */
  var.dmem     = 25;                   /* memory access time */
  var.dmstat   = 5;                    /* memory status time */

  for (i=0; i<2; i++)
  {
    printf("\n\n\n Cost Delay Analysis of the DLX Design \n");
    printf(" No pipelining, hardwired control, no interrupt\n");
    printf(" Technology:  %s, ", name_techno());

    strcpy(my_str,"carry look ahead");
    printf(" Adder type:  %s\n", my_str);
    printf(
      " width n = %d,  Reg-number = %d,  Mem status time = %d\n",
      var.n,var.A,var.dmstat);

/* ------------------------------------------------------------- */
/* Cost of the DP environments and the control                   */
/* ------------------------------------------------------------- */

    forall_line();                     /* cost analysis */
    forall_header("cost","");
    forall_print("ALUenv",caluenv);
    forall_print("SHenv",cshenv);
    forall_print("IRenv",cirenv);
    forall_print("GPRenv",cgpr);
    forall_print("PCenv",cpcenv);
    forall_print("Menv",cmenv);
    forall_line();

    forall_print("DP",cDP);
    forall_print("Control",cCONenv);
    forall_line();
    forall_print("DLX",cDLX);
    forall_print("CON/DLX [%]",cCONrel);

    forall_line();
    forall_print("GPRenv",cgpr);
    forall_print("DP",cDP);
    forall_print("DLX",cDLX);
    forall_line();
    forall_print("GPR/DP [%]",cGPRrel1);
    forall_print("GPR/DLX [%]",cGPRrel2);
    forall_line();
```

```
/* ------------------------------------------------------------ */
/*    Cycle time and delay of DP environments                   */
/* ------------------------------------------------------------ */
    forall_line();                      /* cycle time analysis */
    sprintf(my_str,"(dmem = %d)",var.dmem);
    forall_header("delay",my_str);

    forall_print("aCON_Csig",aCON_Csig);
    forall_print("aCON_Mw",aCON_Mw);
    forall_print("aCON_PCce",aCON_PCce);
    forall_print("agpr_aeqz",agpr_aeqz);
    forall_print("aMenv_mstat",aMenv_mstat);
    forall_line();
    forall_print("t_RFwrite",tgpr__RF);
    forall_print("t_RFread",tgpr_RF);
    forall_print("tSH",tSH);
    forall_print("talu_ZR",talu_ZR);
    forall_print("talu_ZC",talu_ZC);
    forall_line();
    forall_print("tDP",tDP);
    forall_print("tCON",tCON);
    forall_line();
    forall_print("tDLX",tDLX);
    forall_print("T_M()",tM);
    forall_line();
    var.dmem   = 10;                     /* memory access time */
    forall_print("T_M(10)",tM);
    var.dmem   = 50;                     /* memory access time */
    forall_print("T_M(50)",tM);
    forall_line();
    var.dmem   = 25;                     /* memory access time */

    init_techno(VENUS); /* select technology */
  }

  forall_line();
  /* ------------------------------------------------------------ */
  /*  CPI independent                                             */
  /* ------------------------------------------------------------ */
  printf("Computing the EQ-values for the DLX designs \n");
  printf("non-pipelined, no interrupts, hardwired control \n");
  for (i=0; i<2; i++)
  { set(0,i);
    forall_line();
    printf(" Technology: %s \n", name_techno());
    forall_line();
    for (x=0; x<4; x++)
    { printf("\t \t ");
      for(y=0; y<4; y++)
      { if (x!= y)
```

```
          { set(x, i);
            c1 = cDLX();
            t1 = tDLX();
            set(y, i);
            c2 = cDLX();
            t2 = tDLX();
            res = eval_EQ(c1, t1, c2, t2);
          }
          else
            res = 0.0;
          printf("%.3f  ",res);
      }
      printf("\n");
  }
}


/* ------------------------------------------------------------- */
/*  Quality ratio Q(q, Vi, Mealy)                                */
/* ------------------------------------------------------------- */
forall_line();
printf("\n Moore(un, bin) vs. Mealy(un): Motorola x Venus\n");
forall_line();
for (q=0.0; q<1.01; q+=0.02)
{ printf("%.2f    ", q);
  for(i=0; i<2; i++)
  { /* Motorola vs Venus Technology */
    for(x=0; x<2; x++)
    { set(x, i);
      c1 = cDLX();
      t1 = tDLX();
      set(2, i);
      c2 = cDLX();
      t2 = tDLX();
      res = eval_Q(q, c1, t1, c2, t2);
      printf("%f  ",res);
    }
  }
  printf("  1.0 \n");
  }
}
```

## B.7   DLX with Improved ALU

In chapter 7, we modified the ALU in order to speed up the DLX design. Those design changes only impact a few routines of the architectural file **dlx1.c** of the DLX design. Moreover, we also require new functions for the evaluation of the design variants. Only the new or modified procedures are listed below.

```
/* =========================================================== */
/* dlx2.c       architectural module DLX Trade-off Analysis   */
/*       non-pipelined DLX without interrupt handling          */
/*       ( no interrupt handling, hardwired control)           */
/* =========================================================== */
/*                                                             */
/*  ADDER =                            TEST =                   */
/*   0 :  ripple carry adder            0 : test AU result      */
/*   1 :  carry look-ahead adder        1 : test AU inputs      */
/*   2 :  conditional sum adder         2 : 2-cycle test        */
/* ----------------------------------------------------------- */

/* ----------------------------------------------------------- */
/*      ALU Environment (width: var.n)                         */
/* ----------------------------------------------------------- */
/* --- Arithmetic unit AU (with different adders) ----------- */
int cau()
{ switch(var.ADDER)
  { case 0:
      return var.n*cxor + cRCA(var.n);     /* Ripple Carry */
    case 1:
      return var.n*cxor + cCLA(var.n);     /* Carry look-ahead */
    case 2:
      return var.n*cxor + cCSA(var.n);     /* Conditional Sum */
    default:
      fprintf(stderr,"dlx2.c: cau() unknown adder type\n");
      exit(1);
  }
}

int dau()
{ switch(var.ADDER)
  { case 0:
      return dxor + dRCA(var.n);           /* Ripple Carry */
    case 1:
      return dxor + dCLA(var.n);           /* Carry look-ahead */
    case 2:
      return dxor + dCSA(var.n);           /* Conditional Sum */
    default:
      fprintf(stderr,"dlx2.c: dau() unknown adder type\n");
      exit(1);
  }
}
```

```
/* --- ALU ---------------------------------------------------- */
int calu()
{ int help;

  help = cau() +ccomp() +var.n*(cand +cor +cxor) + 4*var.n*cmux;

  if (var.TEST != 2)
    return help;                        /* 1 cycle condition test */
  else
    return help + (var.n + 4)*cff;   /* 2 cycle condition test */
}

int dALU_zc()                            /* delay of zc */
{
  if (var.TEST != 1)
    return dau() + dcomp();               /* Testing ALU Outputs */
  else                                    /* Testing ALU Inputs */
    return dand + dor + max( dau() + max(dor, dinv),
            dxor + dzero(var.n) + max(dor, dand + dinv));
}

/* --- ALU Environment ------------------------------------------ */
int dalue_c_ZC()                         /* delay of ZC */
{ if (var.TEST != 2 )
    return daluglue() + dALU_zc() + ddriv;     /* 1 cycle test */
  else
    return ddriv;                              /* 2 cycle test */
}

/* ----- Delays: two cycle condition test only -------------- */
int aalueCOND_ZC()                        /* (reg) COND -> ZC  */
{ return dcomp() + ddriv;
}

int dalue_s_COND()                       /* Sbus -> (reg) COND */
{ return dau();
}

int dalue_c_COND()                       /* contr. -> COND */
{ return daluglue() + dau();
}

/* -------------------------------------------------------------- */
/*      Cost and delay of the DLX Data Paths                      */
/* -------------------------------------------------------------- */
/* --- ALU cycles ---------------------------------------------- */
int talu_ZR()
{ return max(aCON_Csig() + dalue_c_ZR(), aRegS() + dalue_s_ZR())
        + dRegw() + dff + delta;
}
```

```
int talu_ZC()
{ if (var.TEST != 2 )
    return max(aCON_Csig() +dalue_c_ZC(), aRegS() +dalue_s_ZC())
            + dRegw() + dff + delta;          /* 1 cycle test */
  else
    return max(aCON_Csig() + dalue_c_ZC(), aalueCOND_ZC())
            + dRegw() + dff + delta;          /* 2 cycle test */
}


int talu_COND()                         /* Update COND register */
{ if (var.TEST == 2 )
    return max(aCON_Csig() + dalue_c_COND(),
              aRegS() + dalue_s_COND()) + dff + delta;
  else
    return 0;                                 /* 1 cycle test */
}

/* --- Cycle time of the data paths ------------------------- */
int tDP()
{ return max( max(tgpr__RF(), tgpr_RF()),
              max(tSH(),  max( max(talu_ZR(), talu_ZC()),
                               max(talu_COND(), tRegce())
                             )
                 )
            );
}

/* ----------------------------------------------------------- */
/*      Print Functions (3 adders and 3 test implementations) */
/* ----------------------------------------------------------- */
void line6()
{
  printf("-------------------------------------------------\
-----------------------------------------\n");
  return;
}

void header6(str1,str2)
char *str1, *str2;
{
  printf("%-14s \t        Test Outputs             Test Inputs\
          2 Cycle Test\n",str1);
  printf("%-14s \t    RCA     CLA     CSA     RCA     CLA\
    CSA     RCA     CLA     CSA\n",str2);
  line6();
  return;
}

void print6(a,b,str,func)
```

```
int a[], b[];
char *str;
int (*func)();
{ int add, test;                          /* loop counters */

  printf("%14s:",str);

  for( test=0; test < 3; test++)
  { if (test == 2)
    { /* 2 cycle condition test: parameter set 2 */
      set_conpar(b[0], b[1], b[2], b[3], b[4], b[5], b[6],
      b[7], b[8], b[9]);
    }
    else
    { /* 1 cycle condition test: parameter set 1 */
      set_conpar(a[0], a[1], a[2], a[3], a[4], a[5], a[6],
      a[7], a[8], a[9]);
    }

    for( add=0; add < 3; add++)
    { var.ADDER = add;
      var.TEST  = test;

      printf("\t%7d",(*func)());
    }
  }
  printf("\n");
  return;
}

/* ------------------------------------------------------------- */
/*      Evaluation of the design                                 */
/* ------------------------------------------------------------- */

void main()
{
  char my_str[200];             /* temporary strings */
  int i;                        /* loop variable */

  /* Control logic parameters: set1 and set2              */
  /* # states, inputs, outputs of the control automaton,  */
  /* monomials (#, max, sum),  signal activity (max, sum) */
  /* fanin (max, sum)                                      */
  int set1[] = {22,14,28,22,10,97,10,93,4,32};
  int set2[] = {24,14,28,22,10,97,10,93,4,34};

  init_techno(MOTOROLA);        /* select technology */

  set_contype( 1, 1);           /* Mealy automaton, binary */
```

```
                                    /* design parameters */
    var.n      = 32;                /* width of the data paths */
    var.A      = 32;                /* number of GP-Registers */
    var.dmem   = 25;                /* memory access time */
    var.dmstat = 5;                 /* memory status time */

    for (i=0; i<2; i++)
    {
      printf("\n\n\n Cost Delay Analysis of the DLX Design \n");
      printf(" No pipelining, hardwired control, no interrupts,\
    Mealy automaton, states coded in binary \n");
      printf(" different adders and test implementations    \n");
      printf(
        " width n = %d,  Reg-number = %d,  Mem status time = %d\n",
        var.n,var.A,var.dmstat);
      line6();
      printf(" Technology:  %s\n", name_techno());

/* ------------------------------------------------------------ */
/* Cost of the DP environments and the control                  */
/* ------------------------------------------------------------ */

      line6();                              /* cost analysis */
      header6("cost","");
      print6(set1,set2,"ALUenv",caluenv);
      print6(set1,set2,"SHenv",cshenv);
      print6(set1,set2,"IRenv",cirenv);
      print6(set1,set2,"GPRenv",cgpr);
      print6(set1,set2,"PCenv",cpcenv);
      print6(set1,set2,"Menv",cmenv);
      line6();
      print6(set1,set2,"DP",cDP);
      print6(set1,set2,"Control",cCONenv);
      line6();
      print6(set1,set2,"DLX",cDLX);
      print6(set1,set2,"CON/DLX [%]",cCONrel);
      line6();

/* ------------------------------------------------------------ */
/*     Cycle time and delay of DP environments                  */
/* ------------------------------------------------------------ */

      line6();                              /* cycle time analysis */
      sprintf(my_str,"(dmem = %d)",var.dmem);
      header6("delay",my_str);

      print6(set1,set2,"aCON_Csig",aCON_Csig);
      print6(set1,set2,"aCON_Mw",aCON_Mw);
      print6(set1,set2,"aCON_PCce",aCON_PCce);
      print6(set1,set2,"agpr_aeqz",agpr_aeqz);
```

```
    print6(set1,set2,"aMenv_mstat",aMenv_mstat);
    line6();
    print6(set1,set2,"t_RFwrite",tgpr__RF);
    print6(set1,set2,"t_RFread",tgpr_RF);
    print6(set1,set2,"tSH",tSH);
    print6(set1,set2,"talu_ZR",talu_ZR);
    print6(set1,set2,"talu_ZC",talu_ZC);
    line6();
    print6(set1,set2,"tDP",tDP);
    print6(set1,set2,"tCON",tCON);
    line6();
    print6(set1,set2,"tDLX",tDLX);
    print6(set1,set2,"T_M()",tM);
    var.dmem   = 10;                          /* memory access time */
    print6(set1,set2,"T_M(10)",tM);
    line6();
    var.dmem   = 25;                          /* memory access time */

    init_techno(VENUS); /* select technology */
  }
}
```

## B.8   DLX with Interrupt Handling

In chapter 8, we modified the original DLX design to enable interrupt handling. The design change only impacts a few environments of the DLX. Thus, only a few routines of section B.6 need to be modified. Those routines are listed below.

```c
/* ============================================================ */
/* dlx3.c          architectural module    non-pipelined DLX    */
/*      with interrupt handling, hardwired control, CLA adder */
/* ============================================================ */

/* ------------------------------------------------------------ */
/*    Control Environment (1): Control Signals                  */
/* ------------------------------------------------------------ */
/* --- SPR control circuit (ce and oe for SPRenv) ------------ */
int cSPRcon()
{ return  cdecf(3) + 10*cand + 10*cor;
}

int aSPRcon()
{ return max(ddecf(3), aCON_Csig()) + dand + dor;
}

/* --- Cause Processing Circuit ------------------------------ */
int cCaPro()
{ return 32*cor + 25*cand;
}

int aCaPro()
{ return 5*dor + dand;
}

/* ------------------------------------------------------------ */
/*      Further DP Environments                                 */
/* ------------------------------------------------------------ */
/* --- Memory Environment ------------------------------------ */
int cmenv()
{ return 2 * var.n * cff +var.n * cmux +2 * var.n * cdriv +cor;
}

int dmenv_S1()
{ return ddriv;
}

/* ------------------------------------------------------------ */
/*      SPR Environment                                         */
/* ------------------------------------------------------------ */
int cSPRenv()
{ return 5*var.n*cff + 5*var.n*cdriv + 2*var.n*cmux;
}
```

```
int aSPR__D()
{ return aSPRcon() + ddriv;
}

int dSPR_D()
{ return dmux;
}

int dSPR_MCA()
{ return 0;}


/* --- SPR intern cycle ------------------------------------- */
int tSRsel_SR_SR()
{ return max(aSPRcon(), aCON_Csig() +2*dmux) + dff + delta;
}



/* ------------------------------------------------------------ */
/*      Cost and delay of the DLX Data Paths                   */
/* ------------------------------------------------------------ */
int cDP()
{ return caluenv() + cshenv() + cirenv() + cgpr()
        + cpcenv() + cmenv() + cSPRenv();
}

/* --- clocking of registers in the data paths -------------- */
int tRegce()
{ return aCON_Csig() + dff + delta;
}

/* --- Cycle through the SPRenv ----------------------------- */
/* --- Cycle  in SPR: MCA -> ECA ---------------------------- */
int tSPR_MCA()
{ return aCaPro() + dSPR_MCA() + dff + delta;
}

/* --- Reading out an SPR register -------------------------- */
int tSPR_SR_D()
{ return aSPR__D() + dRegw() + dff + delta;
}

/* --- Cycle time of the data paths ------------------------- */
int tDP()
{ return max( max( max(tSRsel_SR_SR(),
                       max(tSPR_SR_D(), tSPR_MCA()))),
                   max(tgpr__RF(), tgpr_RF()))),
              max(tSH(),
                  max( max(talu_ZR(), talu_ZC()),
                       tRegce()))));
```

```
}

/* --- Memory cycle time tM -------------------------------- */
int aMenv_nready()
{ return max(aMenv_mstat(), aCON_mis()) + dor;
}


/* -------------------------------------------------------- */
/*    Control Environment (2): Next State Computation     */
/* -------------------------------------------------------- */
/* --- accumulated delay: inputs In of control automaton ---- */
int aIn()
{ return max( aCaPro(), max(agpr_aeqz(), aMenv_nready()));
}


/* -------------------------------------------------------- */
/*    Composing the CA environment                         */
/* -------------------------------------------------------- */
/* --- Generation of event signals ------------------------- */
int cevent()
{ return cILL() + 5*cand + cor;}

int aevent()
{ return max( max(aILL(),
                  aMenv_mstat() + dand),
              max( max(aCON_Csig() + dor,
                       aCON_mis() ) + dand,
                   max(aCON_Csig() + dand,
                       aalu_ovf()) + dand
                 )
            );
}

int cCAenv()
{ return var.n*cff + 7*cmux + 7*cor + cevent();
}

int tCAenv()
{ return max(aevent(), aCON_Csig() + dmux) + dor +dff + delta;
}


/* ------ Control Unit ------------------------------------- */
int cCONenv()
{ return cauto() + cMC() + cSPRcon() + cCaPro() + cCAenv();
}

int tCON()
{ return max(tauto(), tCAenv());
}
```

```
/* ------------------------------------------------------------ */
/*      Evaluation of the design                                */
/* ------------------------------------------------------------ */

void main()
{
  char my_str[200];            /* temporary strings */
  int i, x, y;                 /* loop variable */
  int c1, c2, t1, t2;          /* cost, time values */
  double res, q;               /* result, quality parameter */

  init_techno(MOTOROLA);       /* select technology */

                               /* Control logic parameters */
  set_conpar(   32,            /* states of state machine */
             17, 39,           /* input/output: ctr automaton*/
             36, 12, 174,      /* monomials (#, max, sum) */
             15, 137,          /* signal activity (max, sum) */
             6, 52);           /* fanin (max, sum) */

                               /* design parameters */
  var.n       = 32;            /* width of the data paths */
  var.A       = 32;            /* number of GP-Registers */
  var.dmem    = 25;            /* memory access time */
  var.dmstat  = 5;             /* memory status time */

  for (i=0; i<2; i++)
  {
    printf("\n\n\n Cost Delay Analysis of the DLX Design \n");
    printf(" No pipelining, hardwired control, interrupts \n");
    printf(" Technology:  %s, ", name_techno());

    strcpy(my_str,"carry look ahead");
    printf(" Adder type:  %s\n", my_str);
    printf(" width n = %d, Reg-no = %d, Mem status time = %d\n",
           var.n,var.A,var.dmstat);

/* ------------------------------------------------------------ */
/* Cost of the DP environments and the control                  */
/* ------------------------------------------------------------ */

    forall_line();              /* cost analysis */
    forall_header("cost","");
    forall_print("ALUenv",caluenv);
    forall_print("SHenv",cshenv);
    forall_print("IRenv",cirenv);
    forall_print("GPRenv",cgpr);
    forall_print("PCenv",cpcenv);
    forall_print("Menv",cmenv);
```

```
      forall_print("SPRenv",cSPRenv);
      forall_line();

      forall_print("DP",cDP);
      forall_print("Control",cCONenv);
      forall_line();
      forall_print("DLX",cDLX);
      forall_print("CON/DLX [%]",cCONrel);

      forall_line();
      forall_print("GPRenv",cgpr);
      forall_print("DP",cDP);
      forall_print("DLX",cDLX);
      forall_line();
      forall_print("GPR/DP [%]",cGPRrel1);
      forall_print("GPR/DLX [%]",cGPRrel2);
      forall_line();

/* ------------------------------------------------------------ */
/*    Cycle time and delay of DP environments                   */
/* ------------------------------------------------------------ */
      forall_line();                      /* cycle time analysis */
      sprintf(my_str,"(dmem = %d)",var.dmem);
      forall_header("delay",my_str);

      forall_print("aCON_Csig",aCON_Csig);
      forall_print("aCON_Mw",aCON_Mw);
      forall_print("aSPRcon",aSPRcon);
      forall_print("aCaPro",aCaPro);
      forall_print("agpr_aeqz",agpr_aeqz);
      forall_print("aMenv_nready",aMenv_nready);
      forall_print("aalu_ovf",aalu_ovf);
      forall_line();
      forall_print("t_RFwrite",tgpr__RF);
      forall_print("t_RFread",tgpr_RF);
      forall_print("tSH",tSH);
      forall_print("talu_ZR",talu_ZR);
      forall_print("talu_ZC",talu_ZC);
      forall_print("tSRsel_SR",tSRsel_SR_SR);
      forall_print("tSPR_SR_D",tSPR_SR_D);
      forall_print("tSPR_MCA",tSPR_MCA);
      forall_line();
      forall_print("tDP",tDP);
      forall_print("tauto",tauto);
      forall_print("tCAenv",tCAenv);
      forall_print("tCON",tCON);
      forall_line();
      forall_print("tDLX",tDLX);
      forall_print("T_M()",tM);
      forall_line();
```

```
    var.dmem  = 10;                      /* memory access time */
    forall_print("T_M(10)",tM);
    var.dmem  = 50;                      /* memory access time */
    forall_print("T_M(50)",tM);
    forall_line();
    var.dmem  = 25;                      /* memory access time */

    init_techno(VENUS); /* select technology */
}

forall_line();
/* ------------------------------------------------------------- */
/*  CPI independent                                              */
/* ------------------------------------------------------------- */
printf("Computing the EQ-values for the DLX designs \n");
printf("non-pipelined, with interrupts, hardwired control\n");
for (i=0; i<2; i++)
{ set(0,i);
  forall_line();
  printf(" Technology: %s \n", name_techno());
  forall_line();
  for (x=0; x<4; x++)
  { printf("\t \t ");
    for(y=0; y<4; y++)
    { if (x!= y)
      { set(x, i);
        c1 = cDLX();
        t1 = tDLX();
        set(y, i);
        c2 = cDLX();
        t2 = tDLX();
        res = eval_EQ(c1, t1, c2, t2);
      }
      else
        res = 0.0;
      printf("%.3f  ",res);
    }
    printf("\n");
  }
}


/* ------------------------------------------------------------- */
/*  Quality ratio Q(q, Vi, Mealy)                               */
/* ------------------------------------------------------------- */
forall_line();
printf("\n Moore(un,bin) vs. Mealy(bin): Motorola x Venus\n");
forall_line();
for (q=0.0; q<1.01; q+=0.02)
{ printf("%.2f   ", q);
  for(i=0; i<2; i++)
```

```
{ /* Motorola vs Venus Technology */
  for(x=0; x<2; x++)
  { set(x, i);
    c1 = cDLX();
    t1 = tDLX();
    set(3, i);
    c2 = cDLX();
    t2 = tDLX();
    res = eval_Q(q, c1, t1, c2, t2);
    printf("%f  ",res);
  }
}
printf("  1.0 \n");
}
}
```

# B.9   DLX with Microprogrammed Control

In chapter 9, we replaced the hardwired control by six versions of microprogrammed control. That only impacts directly the cost and the delay of the
control automaton, but the hardwired implementation of the control automaton
is covered in the service module hardctr.

Thus, a new service module microctr is required for the microcoded control.
Its header and code file are listed in sections B.9.1 and B.9.2. The architectural
file of the DLX with interrupt handling (section B.8) can almost literally be
copied, but the evaluation part must be adapted. This part of the new DLX
module is listed below.

```
/* ============================================================ */
/* dlx4.c        architectural module    non-pipelined DLX     */
/*        no interrupt handling, microcoded control, CLA adder */
/* ============================================================ */
/* ------------------------------------------------------------ */
/*     Evaluation of the design                                 */
/* ------------------------------------------------------------ */

void main()
{
  char my_str[200];          /* temporary strings */
  int i, x, y;               /* loop variable */
  int c1, c2, ch, t1, t2, th; /* cost, time values */
  double res, fac;           /* result, scaling of ROM cost*/

                             /* Cost, delay of the best DLX
                                with hardwired control      */
  static int MCcost = 15087; /* cost DLX, Motorola */
  static int MCtc   =    80; /* cycle time, Motorola */
  static int VEcost = 33480; /* cost DLX, Venus */
  static int VEtc   =    49; /* cycle time, Venus */

  init_techno(MOTOROLA);     /* select technology */

                             /* Control logic parameters */
  set_micropar( 32,          /* states of state machine */
                54, 29,      /* output (decoded, encoded) */
                12,  7);     /* decode table (full, opt.) */

                             /* design parameters */
  var.n      = 32;           /* width of the data paths */
  var.A      = 32;           /* number of GP-Registers */
  var.dmem   = 25;           /* memory access time */
  var.dmstat = 5;            /* memory status time */

  for (i=0; i<2; i++)
  {
    printf("\n\n\n Cost Delay Analysis of the DLX Design \n");
```

```
    printf(" No pipelining, microcoded control, interrupts \n");
    printf(" Technology:  %s, ", name_techno());

    strcpy(my_str,"carry look ahead");
    printf(" Adder type:  %s\n", my_str);
    printf(" width n = %d, Reg-no = %d, Mem status time = %d\n",
           var.n,var.A,var.dmstat);

/* ------------------------------------------------------------ */
/* Cost of the DP environments and the control                 */
/* ------------------------------------------------------------ */

    sixall_line();                    /* cost analysis */
    sixall_header("cost","");
    sixall_print("ALUenv",caluenv);
    sixall_print("SHenv",cshenv);
    sixall_print("IRenv",cirenv);
    sixall_print("GPRenv",cgpr);
    sixall_print("PCenv",cpcenv);
    sixall_print("Menv",cmenv);
    sixall_print("SPRenv",cSPRenv);
    sixall_line();

    sixall_print("DP",cDP);
    sixall_print("Control",cCONenv);
    sixall_line();
    sixall_print("DLX",cDLX);
    sixall_print("CON/DLX [%]",cCONrel);

    sixall_line();
    sixall_print("GPRenv",cgpr);
    sixall_print("DP",cDP);
    sixall_print("DLX",cDLX);
    sixall_line();
    sixall_print("GPR/DP [%]",cGPRrel1);
    sixall_print("GPR/DLX [%]",cGPRrel2);
    sixall_line();

/* ------------------------------------------------------------ */
/*    Cycle time and delay of DP environments                  */
/* ------------------------------------------------------------ */
    sixall_line();                    /* cycle time analysis */
    sprintf(my_str,"(dmem = %d)",var.dmem);
    sixall_header("delay",my_str);

    sixall_print("aCON_Csig",aCON_Csig);
    sixall_print("aCON_Mw",aCON_Mw);
    sixall_print("aSPRcon",aSPRcon);
    sixall_print("aCaPro",aCaPro);
    sixall_print("agpr_aeqz",agpr_aeqz);
```

```
    sixall_print("aMenv_nready",aMenv_nready);
    sixall_print("aalu_ovf",aalu_ovf);
    sixall_line();
    sixall_print("t_RFwrite",tgpr__RF);
    sixall_print("t_RFread",tgpr_RF);
    sixall_print("tSH",tSH);
    sixall_print("talu_ZR",talu_ZR);
    sixall_print("talu_ZC",talu_ZC);
    sixall_print("tSRsel_SR",tSRsel_SR_SR);
    sixall_print("tSPR_SR_D",tSPR_SR_D);
    sixall_print("tSPR_MCA",tSPR_MCA);
    sixall_line();
    sixall_print("tDP",tDP);
    sixall_print("tauto",tauto);
    sixall_print("tCAenv",tCAenv);
    sixall_print("tCON",tCON);
    sixall_line();
    sixall_print("tDLX",tDLX);
    sixall_print("T_M()",tM);
    sixall_line();
    var.dmem    = 10;                      /* memory access time */
    sixall_print("T_M(10)",tM);
    var.dmem    = 50;                      /* memory access time */
    sixall_print("T_M(50)",tM);
    sixall_line();
    var.dmem    = 25;                      /* memory access time */
    sixall_line();

/* ------------------------------------------------------------ */
/*   Parameter study: T(auto, dmstat)                           */
/* ------------------------------------------------------------ */
    sixall_line();
    sixall_header("tauto","dmstat =");
    y = var.dmstat;
    for (x = 0; x < 55; x+=10)
    { var.dmstat = x;
      sprintf(my_str," %d ",var.dmstat);
      sixall_print(my_str,tauto);
    }
    var.dmstat = 5;

    init_techno(VENUS); /* select technology */
  }

sixall_line();
/* ------------------------------------------------------------ */
/*  Standard scaling of ROM cost: EQ() for Encoding             */
/* ------------------------------------------------------------ */
printf("Computing the EQ-values for the DLX designs \n");
printf("non-pipelined, with interrupt, microcoded control\n");
```

```
    printf("(Table optimal): unary, binary-pre, binary-post \n");
    for (i=0; i<2; i++)
    { if (i == 0)
        init_techno(MOTOROLA);
      else
        init_techno(VENUS);

      sixall_line();
      printf(" Technology: %s \n", name_techno());
      sixall_line();
      for (x=0; x<3; x++)
      { printf("\t \t ");
        for(y=0; y<3; y++)
        { if (x!= y)
          { set_microtype(1, x);
            c1 = cDLX();
            t1 = tDLX();
            set_microtype(1, y);
            c2 = cDLX();
            t2 = tDLX();
            res = eval_EQ(c1, t1, c2, t2);
          }
          else
            res = 0.0;
          printf("%.3f  ",res);
        }
        printf("\n");
      }
    }

/* ------------------------------------------------------------ */
/*  Variable scaling of the ROM cost                            */
/* ------------------------------------------------------------ */
sixall_line();
printf("          | MOTOROLA       | VENUS  \n");
printf("             EQ        Cost       EQ        Cost \n");
printf(" rfac    V1,V2      V1,HW     V1,V2      V1,HW \n");
sixall_line();
for (fac=1.0; fac<40.2; fac+=0.20)
{ printf("%5.2f    ",fac);
  for(i=0; i<2; i++)
  { /* Motorola vs Venus Technology */
    if (i == 0)
    { init_techno(MOTOROLA);
      ch = MCcost;
      th = MCtc;
    }
    else
    { init_techno(VENUS);
      ch = VEcost;
```

```
        th = VEtc;
    }

    romfac = fac;                       /* scaling of ROM cost */

    set_microtype(1, 1);                /* Opt. Table, pre-binary */
    c1 = cDLX();
    t1 = tDLX();

    set_microtype(1, 2);                /* Opt. Table, post-bin. */
    c2 = cDLX();
    t2 = tDLX();

    res = eval_EQ(c1, t1, c2, t2);      /* EQ(V1,V2) */
    printf("%f  ",res);

    res = (1.0 * c1)/(1.0 * ch);        /* Cost(V1,Hardwired) */
    printf("%f  ",res);

  }
  printf("  1.0 \n");
 }
}
```

## B.9.1   Header File of the Microcoded Control

```
/* ************************************************************
** microctr.h          microcoded control module          **
*************************************************************/
/* Table =                  Coding =                        */
/* 0 : full decode table    0 : unary coded micro-instruction */
/* 1 : optimized decode     1 : bin, decoders infront micro-IR */
/*                          2 : bin, decoders behind micro-IR */
/* ---------------------------------------------------------- */

#ifndef MICROC_H
#define MICROC_H

extern int set_microtype();     /* set type of microcoded ctr */
                                /* Table, Coding */
extern int set_micropar();      /* set automaton parameters */
                                /* states, alpha, widtha,
                                   fsize, osize */
extern int aOut();              /* accum. delay of outputs */
extern int cauto();             /* cost of microcoded ctr */
extern int tauto();             /* cycle time of micro-ctr */

/* aCaPro(), agpr_aeqz();       to be defined in machine spec */
/* aMenv_nready();              accum. delay of inputs */
```

```
                                /* circuit ILL detects illegal
                                   instructions */
extern int cILL();              /* cost of ILL */
extern int aILL();              /* accum. delay of ILL */

extern void sixall_line();      /* print horizontal line */
extern void sixall_header();    /* print two line header */
extern void sixall_print();     /* evaluate function, print */

struct micro_parameters {
  int states, zeta;     /* #States of automaton, ld(states) */
  int alpha, widtha;    /* output: width (decoded, encoded) */
  int fsize, osize;     /* width: decode table (full, optimal */
  int Table, Coding;    /* ctr type: decode table, outputs */
};

#endif MICROC_H
```

## B.9.2   Code File of the Microcoded Control

```
/* ***************************************************************
** microctr.c            microcoded control module          **
***************************************************************/

#include <stdio.h>
#include "techno.h"
#include "macro.h"
#include "microctr.h"

static struct                   /* private control parameters */
        micro_parameters con =
        {0,0, 0,0        ,      /* must be set by set_micro.. */
         0,0, 0,0};

int set_microtype(t,c)          /* set type of microcoded ctr */
int t,c;                        /* Table, Coding */
{
  con.Table = t;
  con.Coding = c;
  return 0;
}

int set_micropar(s,al,a,fs,os)
int s,al,a,fs,os;
                                /* set micro ctrl parameters */
{
  con.states   = s;            /* #states */
  con.zeta     = ld(s);
  con.alpha    = al;           /* #outputs (decoded) */
  con.widtha   = a;            /* #outputs (encoded) */
```

```
  con.fsize    = fs;             /* size decode table, full */
  con.osize    = os;             /* size decode table, optimal */
  return 0;
}


/* --- Decode Unit: decoding the micro-instruction ---------- */
int cMuDec()
{ return 5*cdecf(3) + 4*cdecf(2) + cdecf(1);
}

int dMuDec()
{ return ddecf(3);
}

/* --- Accumulated delay: Csig and microcondition ----------- */
int aOut()
{ if (con.Coding == 2)
    return dMuDec();
  else
    return 0;
}

/* --- Accum. delay: inputs of the microcoded automaton ----- */
extern int aCaPro(), aMenv_nready(), agpr_aeqz();

/* --- Decode Lookup Table ---------------------------------- */
int cDecTab()
{ if (con.Table == 1)
    return  crom(power2(con.osize), con.zeta)
                + czero(con.zeta) + con.zeta * cmux;
  else
    return crom(power2(con.fsize), con.zeta);
}

int aDecTab()
{ if (con.Table == 1)
    return drom(power2(con.osize), con.zeta) + dzero(con.zeta)
           + dmux;
  else
    return drom(power2(con.fsize), con.zeta);
}

/* --- Address Selection Unit ------------------------------- */
int cAdrSel()
{ return cDecTab() + 6*con.zeta*cmux + 6*cand + 6*cor + cxnor;}

int aAdrSel()
{ int d1, d2, d3, d4;
```

```
    d1 = max( max(aDecTab(), aOut() + dmux), dinc(con.zeta))
         + 3*dmux;
    d2 = max(aOut(), aCaPro()) + dand + dor + dmux
         + max(dmux, dor);
    d3 = aMenv_nready() + 4*dmux;
    d4 = max(aOut() + 3*dor, agpr_aeqz() + dxnor + dor)
         + dand + 3*dmux;

    return max( max(d1, d2), max(d3, d4));
}

/* --- Cost of control automaton ---------------------------- */
int cauto()
{ int cost;
  cost = con.zeta *(cff +cand) +cinv +cinc(con.zeta) +cAdrSel();

  switch(con.Coding)
  { case 0:
      return  cost +crom(con.states, con.alpha) +con.alpha *cff;
      break;

    case 1:
      return  cost + crom(con.states, con.widtha)
              + con.alpha * cff + cMuDec();
      break;

    case 2:
      return  cost + crom(con.states, con.widtha)
              + con.widtha * cff + cMuDec();
      break;

    default:
      fprintf(stderr,
              "micro.c: cauto() con.Coding out of range\n");
      exit(1);
  }

}

/* --- cycle time of the control automaton ------------------ */
int tauto()
{ int delay;

  delay = delta + dff + dand + aAdrSel();

  switch(con.Coding)
  { case 0:
      return  delay + drom(con.states, con.alpha);
      break;
```

```
    case 1:
      return  delay + drom(con.states, con.widtha) + dMuDec();
      break;

    case 2:
      return  delay + drom(con.states, con.widtha);
      break;

    default:
      fprintf(stderr,
              "micro.c: tauto() con.Coding out of range\n");
      exit(1);
  }

}

/* --- Testing for illegal instruction ---------------------- */
int cILL()
{ return czero(con.zeta) + cand;
}

int aILL()
{ return max( aCON_Csig(), aDecTab() + dzero(con.zeta)) + dand;
}

/* === Print functions ======================================= */
void sixall_line()
{
  printf("-----------------------------------------------------\
------------\n");
  return;
}

void sixall_header(str1,str2)
char *str1, *str2;
{
  printf(
    "%-14s \t    ROM (unary)    binary Pre-IR    bin. Post-IR\n",
    str1);
  printf(
    "%-14s \t    full     opt.     full     opt.     full     opt.\n",
    str2);
  sixall_line();
  return;
}

void sixall_print(str,func)
char *str;
int (*func)();
{ printf("%14s:",str);
```

```
    set_microtype(0, 0); printf("\t%7d",(*func)());
    set_microtype(1, 0); printf("\t%7d",(*func)());
    set_microtype(0, 1); printf("\t%7d",(*func)());
    set_microtype(1, 1); printf("\t%7d",(*func)());
    set_microtype(0, 2); printf("\t%7d",(*func)());
    set_microtype(1, 2); printf("\t%7d\n",(*func)());
    return;
}
```

# Bibliography

[For92]    A. Formella. *Leistung und Güte numerischer Vektorrechnerarchitekturen.* PhD thesis, University of Saarland, Computer Science Department, 1992.

[Grü94]    T. Grün. *Quantitative Analyse von I/O-Architekturen.* PhD thesis, University of Saarland, Computer Science Department, 1994.

[Gwe94]    L. Gwennap. MIPS R10000 uses decoupled architecture. *Microprocessor Report*, 8(14):18–22, 1994.

[Har64]    J. Hartmanis. Loop-free structure of sequential machines. In E.F. Moore, editor, *Sequential Machines.* Addison-Wesley, 1964.

[HNS86]    E. Hörbst, M. Nett, and H. Schwärtzel. *VENUS: Entwurf von VLSI Schaltungen.* Springer, 1986.

[Hot72]    G. Hotz. *Rechenanlagen.* Teubner, 1972.

[Hot74]    G. Hotz. *Schaltkreistheorie.* de Gruyter, 1974.

[HP90]    J.L. Hennessy and D.A. Patterson. *Computer Architecture: A Quantitative Approach.* Morgan Kaufmann Publishers, INC., San Mateo, CA, 1990.

[HP93]    J.L. Hennessy and D.A. Patterson. *The Hardware/Software Interface.* Morgan Kaufmann Publishers, INC., San Mateo, CA, 1993.

[HU79]    J.E. Hopcroft and J.D. Ullman. *Introduction to Automata Theory, Languages, and Computation.* Addison-Wesley, 1979.

[Int89]    *i860 64-Bit Microprocessor Programmer's Reference Manual.* Intel Corporation, Santa Clara, CA, 1989.

[Kel92]    J. Keller. *Zur Realisierbarkeit des PRAM Modelles.* PhD thesis, University of Saarland, Computer Science Department, 1992.

[KH92]    G. Kane and J. Heinrich. *MIPS RISC architecture.* Prentice Hall, 1992.

[KII90]   S. Kotani, A. Inoue, and S. Inanura, T. Hasuo. A 1GOPS 8bit joseph-
          son digital signal processor. In *Proc. IEEE International Solide-State
          Circuits Conference*, pages 148–149, 286, 1990.

[Knu95]   R. Knuth. Quantitative Analyse einer DLX-Architektur mit Pipelin-
          ing. Internal report, University of Saarland, Computer Science De-
          partment, 1995.

[KP95]    J. Keller and W.J. Paul. *Einführung in die Technische Informatik:
          Hardware Design*. Teubner, 1995.

[Mas94]   W. Massonne. *Leistung und Güte von Datenflußrechnern*. PhD thesis,
          University of Saarland, Computer Science Department, 1994.

[Mes94]   H.-P. Messmer. *Pentium*. Addison-Wesley, Bonn, 1994.

[MP90]    S.M. Müller and W.J. Paul. Towards a formal theory of computer
          architecture. In *Proc. PARCELLA 90: Research in Informatics*, vol-
          ume 2, pages 157–169, Berlin, 1990. Akademie–Verlag.

[Mül91]   S.M. Müller. *RISC und CISC: Optimierung und Vergleich von Ar-
          chitekturen*. PhD thesis, University of Saarland, Computer Science
          Department, 1991.

[NB93]    C. Nakata and J. Brock. *H4C Series: Design Reference Guide. CAD,
          0.7 Micron $L_{eff}$*. Motorola Ltd., 1993. Preliminary.

[Sav87]   J.E. Savage. *The Complexity of Computing*. John Wiley & Sons, 1987.

[SBN82]   D.P. Siewiorek, C.G. Bell, and A. Newell. *Computer Structures, Prin-
          ciples & Examples*. McGraw Hill, 1982.

[Sch92]   D. Schmidt. *Leistung und Optimierung von CPU-Architekturen für
          VAX- und IBM/370-Instruktionssätze*. PhD thesis, University of Saar-
          land, Computer Science Department, 1992.

[Sie88]   Siemens München. *VENUS-S Semi-Custom Design System: Zellka-
          talog*, 1988.

[Sla94]   M. Slater. AMD's K5 designed to outrun Pentium. *Microprocessor
          Report*, 8(14):1, 6–11, 1994.

[Spa76]   0. Spaniol. *Arithmetik in Rechenanlagen*. Teubner, 1976.

[WE85]    N. Weste and K. Eshraghian. *Principles of CMOS VLSI Design - A
          System Perspective*. Addison-Wesley, 1985.

[Weg87]   I. Wegener. *The Complexity of Boolean Functions*. John Wiley &
          Sons, 1987.

[WH90]     S.A. Ward and R.H. Halstead Jr. *Computation Structures*. The MIT electrical engineering and computer science series. The MIT Press, Cambridge, Mass., 1990.

[Wil51]    M. V. Wilkes. The best way to design an automatic calculating machine. In *Manchester University Computer Inaugural Conference*. Ferranti Ltd. London, 1951. Reprint in *The Genesis of Microprogramming* in Annuals of the History of Computing, 8(2):116, 1986.

[Wil86]    M. V. Wilkes. The genesis of microprogramming. *Annals of the History of Computing*, 8(2):116–126, 1986.

[WKT51]    F.C. Wiliams, T. Kilburn, and G.C. Tootill. Universal high-speed digital computers: A small scale experimental machine. *IEE Proceedings*, 98(61):13–28, 1951.

[WS53]     M. V. Wilkes and J. B. Stringer. Micro-programming and the design of the control circuits in an electronic digital computer. In *Proc. Cambridge Philosophical Society*, volume 49, pages 230–238, 1953. Reprint in *The Genesis of Microprogramming* in Annuals of the History of Computing, 8(2):116, 1986.

[Yam90]    Kazushige Yamamoto. A subscriber line interface processor for asynchronous tranfer mode switching systems. In *Proc. IEEE International Solide-State Circuits Conference*, pages 28–29, 256, 1990.

# Lecture Notes in Computer Science

For information about Vols. 1–928

please contact your bookseller or Springer-Verlag

# Index

# Springer-Verlag
# and the Environment

We at Springer-Verlag firmly believe that an international science publisher has a special obligation to the environment, and our corporate policies consistently reflect this conviction.

We also expect our business partners – paper mills, printers, packaging manufacturers, etc. – to commit themselves to using environmentally friendly materials and production processes.

The paper in this book is made from low- or no-chlorine pulp and is acid free, in conformance with international standards for paper permanency.